ROLE OF SEDIMENT TRANSPORT IN OPERATION AND MAINTENANCE OF SUPPLY AND DEMAND BASED IRRIGATION CANALS

APPLICATION TO MACHAI MAIRA BRANCH CANALS

ROLE OF SEDIMENT TRANSPORT IN OPERATION AND MAINTENANCE OF SUPPLY AND DEMAND BASED IRRIGATION CANALS

APPLICATION TO MACHAI MAIRA BRANCH CANALS

Thesis
Submitted in fulfilment of the requirements of
the Academic Board of Wageningen University and
the Academic Board of the UNESCO-IHE Institute for Water Education
for the degree of doctor
to be defended in public
on Thursday, 24 February 2011 at 3 p.m.
in Delft, the Netherlands

by

Sarfraz Munir
Born in Narowal, Pakistan

CRC Press/Balkema is an imprint of the Taylor & Francis Group, an informa business

© 2011, Sarfraz Munir

Published by
CRC Press/Balkema
PO Box 447, 2300 AK Leiden, the Netherlands
e-mail: Pub.NL@taylorandfrancis.com
www.crcpress.com - www.taylorandfrancis.co.uk - www.ba.balkema.nl
ISBN 978-0-415-66947-4 (Taylor & Francis Group)
ISBN 978-90-8585-850-8 (Wageningen University)

Table of contents

TABLE OF CONTENTS V

ACKNOWLEDGEMENT XI

SUMMARY XIII

1. INTRODUCTION AND BACKGROUND 1

1.1 BACKGROUND 1
1.2 RATIONALE OF THE STUDY 2
1.3 SCOPE OF THE STUDY 3
1.4 RESEARCH QUESTIONS 3
1.5 RESEARCH HYPOTHESIS 4
1.6 OBJECTIVES 4
1.7 INTRODUCTION OF THE THESIS 4

2. INDUS BASIN IRRIGATION SYSTEM IN PAKISTAN 7

2.1 GENERAL INFORMATION ABOUT PAKISTAN 7
2.2 SOCIO-ECONOMIC SETTINGS OF THE COUNTRY 8
2.3 WATER RESOURCES 8
2.4 INDUS BASIN IRRIGATION SYSTEM 9
2.5 HISTORY OF IRRIGATION IN THE INDUS BASIN 9

3. SEDIMENT TRANSPORT AND FLOW CONTROL IN IRRIGATION
    CANALS 13

3.1. GENERAL INFORMATION ON SEDIMENT TRANSPORT AND MODELLING 13
    3.1.1 Incipient motion 13
    3.1.2 Bed load sediment transport 14
    3.1.3 Suspended sediment transport 15
    3.1.4 Total sediment load 15
3.2. EVALUATION OF SEDIMENT TRANSPORT FORMULAE 16
3.3. SEDIMENT TRANSPORT MODELLING 17
3.4. SEDIMENT TRANSPORT MODELLING IN IRRIGATION CANALS 19
3.5. SEDIMENT MANAGEMENT IN IRRIGATION CANALS 20
    3.5.1 Sediment control at intakes 21
    3.5.2 Sediment diverters (silt excluders) 22
    3.5.3 Sediment ejectors 23
    3.5.4 Settling basins 25
    3.5.5 Sediment control by canal design approach 25
    3.5.6 Operation and maintenance of silt affected irrigation canals 29
3.6. FLOW CONTROL IN IRRIGATION CANALS 30
    3.6.1 Upstream control 30
    3.6.2 Proportional control 30
    3.6.3 Downstream control 30

3.7.   HYDRAULICS OF DOWNSTREAM CONTROL SYSTEM 31
3.8.   DESIGN CRITERIA 34
3.9.   FLOW CONTROL ALGORITHMS 38
    3.9.1. Supervisor Control and Data Acquisition System.........................39
    3.9.2. Proportional Integral Derivative (PID) control............................40

4.   SEDIMENTATION IN TARBELA RESERVOIR AND ITS EFFECTS
    ON PEHURE HIGH LEVEL CANAL...41

4.1.   SEDIMENTATION OF RESERVOIRS IN THE WORLD 41
4.2.   TARBELA DAM 41
4.3.   SEDIMENTATION IN TARBELA RESERVOIR..44
    4.3.1   Water availability at Tarbela Dam 44
    4.3.2   Sediment inflow to Tarbela Reservoir......................................44
    4.3.3   Tarbela Reservoir operation..46
    4.3.4   Sediment trapping in the reservoir and status of storage capacity...........46
    4.3.5   Accumulated sediment deposition in Tarbela Reservoir......................47
4.4.   DEVELOPMENT OF SEDIMENT DELTA IN A RESERVOIR 49
    4.4.1   Siltation process in reservoirs..49
    4.4.2   Formation of a delta 49
    4.4.3   Factors affecting sediment delta movement.............................50
    4.4.4   Reworking of the delta 51
4.5.   ANALYSIS OF THE SEDIMENT DELTA IN TARBELA RESERVOIR...............51
4.6.   PROSPECTS OF SEDIMENT MANAGEMENT...53
4.7.   TARBELA DAM AND PEHURE HIGH LEVEL CANAL 54

5.   HYDRODYNAMIC BEHAVIOUR OF MACHAI MAIRA AND
    PEHURE HIGH LEVEL CANALS 57

5.1.   BACKGROUND 57
5.2.   OBJECTIVES OF THE HYDRODYNAMIC MODELLING 58
5.3.   METHODS AND MATERIALS 58
    5.3.1. Irrigation infrastructure 58
    5.3.2. System operations..62
    5.3.3. Crop Based Irrigation Operations (CBIO)..............................62
    5.3.4. Discharge and water level controls.......................................63
    5.3.5. Secondary offtakes operation...63
    5.3.6. Modelling canal operations..63
    5.3.7. Calibration of the model..64
5.4.   RESULTS AND DISCUSSION..65
    5.4.1. Model calibration and validation 65
    5.4.2. Steady state simulations..66
    5.4.3. Performance of Proportional Integral discharge controllers......................72
    5.4.4. Effect of amount and location of discharge refusal on discharge control....74
    5.4.5. Testing of CBIO schedules...75
    5.4.6. Gate responses..77

6.   SEDIMENT TRANSPORT IN MACHAI AND MAIRA BRANCH
    CANALS 79

6.1.   METHODOLOGY OF FIELD DATA COLLECTION 80
    6.1.1. Measurement methods 81
    6.1.2. Fieldwork arrangement 87
6.2.   IRRIGATION WATER SUPPLY TO USC-PHLC SYSTEM 88
6.3.   WATER DISTRIBUTION TO SECONDARY OFFTAKES 89
6.4.   SEDIMENT INFLOW TO THE SYSTEM 91
6.5.   RESULTS FROM MASS BALANCE STUDIES 94
    6.5.1. Mass balance in July 2007 94
    6.5.2. Mass balance in December 2007 97
    6.5.3. Mass balance in July 2008 99
6.6.   OFFTAKES SEDIMENT WITHDRAWAL 102
6.7.   MORPHOLOGICAL CHANGES IN THE CANALS 106
    6.7.1. Cross-sectional survey in January 2007 106
    6.7.2. Cross-sectional survey in July 2007 106
    6.7.3. Cross-sectional survey in January 2008 107

7.   CANAL OPERATION AND SEDIMENT TRANSPORT 109

7.1.   BACKGROUND 109
7.2.   SIMULATION OF IRRIGATION CANALS (SIC) MODEL 109
    7.2.1. Topography module 109
    7.2.2. Steady flow module 110
    7.2.3. Unsteady flow module 111
    7.2.4. Sediment module 112
7.3.   SEDIMENT TRANSPORT MODEL SET UP 117
    7.3.1. Sensitivity analysis 118
    7.3.2. Comparison between sediment transport predictors 118
    7.3.3. Model calibration and validation 120
7.4.   SCENARIO SIMULATION 122
    7.4.1. Simulations under existing conditions of water and sediment discharge ... 123
    7.4.2. Sediment transport under design discharges with existing sediment
           concentration 123
    7.4.3. Sediment transport under CBIO with existing sediment concentration 124
7.5.   SEDIMENT TRANSPORT IN PHLC 125
    7.5.1. Sediment transport capacity in PHLC 126
    7.5.2. Sediment transport capacity in the canals downstream of RD 242 127
7.6.   SCENARIO SIMULATIONS WITH TARBELA EFFECT 129
    7.6.1. Sediment transport at full supply discharge 130
    7.6.2. Sediment transport at existing discharge conditions 132
    7.6.3. Sediment transport under CBIO 134
7.7.   EFFECT OF FLOW CONTROL ON SEDIMENT TRANSPORT 136
7.8.   SEDIMENT CONCENTRATION ARRIVING AT RD 242 FROM THE PHLC 137
7.9.   COMBINED SEDIMENT CONCENTRATION AT CONFLUENCE FOR CBIO 138
7.10.   SEDIMENT TRANSPORT DOWNSTREAM OF RD 242 WITH COMBINED
        SEDIMENT INFLOW FROM MACHAI BRANCH CANAL AND PHLC 139
    7.10.1. At design discharge 139
    7.10.2. At existing discharge 141
    7.10.3. At Crop Based Irrigation Operations 142

8.   DEVELOPMENT OF DOWNSTREAM CONTROL COMPONENT IN
            SETRIC MODEL 143

    8.1.   BACKGROUND 143
    8.2.   RATIONALE 144
    8.3.   SETRIC MODEL 145
        8.3.1. Water flow calculations 145
        8.3.2. Roughness calculations 146
        8.3.3. Determination of roughness on the side walls 147
        8.3.4. Galappatti's depth integrated model 148
        8.3.5. Separation of bed and suspended load 149
        8.3.6. Concentration downstream of inflow and outflow points 150
        8.3.7. Morphological changes in the bed 150
        8.3.8. Boundary conditions 152
    8.4.   IMPROVEMENTS IN THE SETRIC MODEL 152
        8.4.1. AVIS and AVIO gates 153
        8.4.2. Gate index of AVIS/AVIO gates 153
        8.4.3. Hydraulics of AVIS/AVIO gates 154
        8.4.4. Discharge computation 155
    8.5.   COMPUTATION PROCEDURE IN SETRIC MODEL FOR DOWNSTREAM
            CONTROL 156
    8.6.   APPLICATION OF THE SETRIC MODEL TO AUTOMATICALLY
            DOWNSTREAM CONTROLLED IRRIGATION CANAL 157
        8.6.1. Calibration and validation of the model 158
    8.7.   FLOW SIMULATIONS 158
        8.7.1. Flow simulation under design flow conditions 158
        8.7.2. Flow simulation at 50% of full supply discharge 159
        8.7.3. Flow simulation at 75% of full supply discharge 160
    8.8.   SEDIMENT TRANSPORT SIMULATIONS 161
        8.8.1. Calibration and validation of SETRIC Model for sediment transport 161
        8.8.2. Sediment inflow to the irrigation canal under study 162
        8.8.3. Under design discharge with existing sediment inflow 162
        8.8.4. Under existing water and sediment inflow 162
        8.8.5. Under CBIO and existing sediment inflow 163

9.   MANAGING SEDIMENT TRANSPORT 165

    9.1.   TRADITIONAL APPROACH OF SEDIMENT MANAGEMENT IN INDUS BASIN
            IRRIGATION SYSTEM 165
    9.2.   DESILTING OF IRRIGATION CANALS 167
    9.3.   EFFECT OF SEDIMENT TRANSPORT ON UPSTREAM CONTROLLED
            IRRIGATION CANALS 168
    9.4.   EFFECT OF SEDIMENT TRANSPORT ON THE HYDRAULIC PERFORMANCE
            OF DOWNSTREAM CONTROLLED IRRIGATION CANAL 169
        9.4.1. Effect of sediment deposition at the automatic flow releases 169
        9.4.2. Effect of secondary offtakes operation on sediment transport capacity 172
    9.5.   EFFECT OF DIFFERENT OPERATION SCHEMES ON SEDIMENT TRANSPORT 173
        9.5.1. Effects of design discharge 173

9.5.2. *Effects of existing discharge* *174*
9.5.3. *Effect of different options of CBIO on sediment transport* *176*
9.6.   MANAGEMENT OPTIONS 178
9.6.1. *Operation under different discharge conditions* *178*
9.6.2. *Sediment transport under CBIO with increased sediment discharge from
        the Tarbela Reservoir* *184*
9.6.3. *Target water level and sediment transport* *186*
9.6.4. *Decrement setting and AVIS gates' response* *187*
9.6.5. *Grouping and clustering of offtakes* *188*
9.6.6. *Offtakes close to cross regulators* *189*

10.   EVALUATION 191
10.1. SEDIMENTATION IN TARBELA RESERVOIR AND PHLC 191
10.2. HYDRODYNAMIC BEHAVIOUR OF THE IRRIGATION CANALS 191
10.3. FLOW AND SEDIMENT TRANSPORT IN THE IRRIGATION CANALS 192
10.4. SEDIMENT TRANSPORT MODELLING 194
10.5. DEVELOPMENT OF DOWNSTREAM CONTROL COMPONENT IN SETRIC
       MODEL 196
10.6. MANAGING SEDIMENT TRANSPORT 196
10.7. CONCLUSIONS AND WAY FORWARD 198

REFERENCES 199

LIST OF SYMBOLS 207

ACRONYMS 211

APPENDICES 213

A.   SEDIMENT TRANSPORT RELATIONSHIPS UNDER EQUILIBRIUM
      CONDITIONS 213
B.   THE SAINT-VENANT EQUATIONS AND THEIR SOLUTION 221
C.   ONE DIMENSIONAL CONVECTION DIFFUSION EQUATION AND ITS
      SOLUTION IN UNSTEADY STATE FLOW CONDITIONS 227
D.   MODIFIED EINSTEIN PROCEDURE FOR COMPUTATION OF TOTAL
      SEDIMENT LOAD FROM FIELD MEASUREMENTS 233
E.   CROP WATER REQUIREMENTS AND CROP BASED IRRIGATION
      OPERATIONS 241
F.   SEDIMENT INFLOW AT MACHAI BRANCH HEADWORKS 251
G.   SEDIMENT INFLOW AT RD 242 255
H.   RESULTS OF EQUILIBRIUM MEASUREMENTS 259

SAMENVATTING 263

ABOUT THE AUTHOR 269

ACKNOWLEDGEMENT

I would like to thank my promoter Prof. E. Schultz, PhD, MSc for his kind supervision, academic advice, and continuous support during my PhD study period. I would also like to thank my mentor F.X. Suryadi, PhD, MSc who provided me guidance and with whom I had fruitful discussions on the day to day problems and issues related to my study. I also thank Dr. Chu Thai Hoanh for his supervision of my study.

Many thanks go to Mr. Khalid Mohtadullah and Mr. Abdul Hakeem Khan, the country directors of the International Water Management Institute (IWMI) for their wholehearted support and guidance during my stay in Pakistan. I am grateful to Dr. Hugh Turral (IWMI) and Dr. Zhong Ping Zhu (IWMI) who encouraged me to take up this PhD study.

I would like to thank Mr. Krishna P. Paudel for sharing nice ideas and having nice discussions on the topic. I would also like to thank Mr. Paul Ankum, Dr. Peter Jules van Overloop and Dr. Arnold Lobbrecht for the discussions with them on the topic of canal automation.

I am special thankful to Pierre-Olivier Malaterre, Gilles Belaud and Litrico Xavier for their time to time support to me in learning and applying various modules of the Simulation of Irrigation Canals (SIC) Model.

I am thankful to the International Sedimentation Research Institute of Pakistan (ISRIP) for their help in water and sediment field measurements. I immensely benefited from their knowledge, experience and literature on sediment transport in the Indus Basin Irrigation System (IBIS). I really enjoyed their company during field measurements. I also extend my thanks to Mr. Noor ul Amin for his continuous support during field data collection. I am grateful to the Water and Power Development Authority (WAPDA), Pakistan for sharing information on Indus River flows and Tarbela Reservoir.

I would like to express my deep sense of gratitude to the Khyber Pakhtunkhwa Department of Irrigation for data sharing on canal flows, their logistic support and cooperation during the field measurements.

I am thankful to the IWMI and UNESCO-IHE for the financial and logistic support they provided for this study.

I extend my thanks to Jolanda Boots and staff members of the Core Land and Water Development who helped me during my study.

I really enjoyed the company of my friends Naveed Alam, Shahid Ali, Muhammad Jehanzeb Masud Cheema, Faisal Karim, S. H. Sandilo, Bilal Ahmad from Delft University of Technology, Ilyas Masih here at UNESCO-IHE and Saim Muhammad from Utrecht. It would have been much more difficult for me to live and work away from my family, without their company and support.

Last but not the least, I am thankful to my family whose support and affection encouraged me to complete this study. I am especially thankful to the cooperation, continuous support and endless patience of my wife and children during this study period.

Summary

Like in many emerging and least developed countries, agriculture is vital for Pakistan's national economy. It contributes 21% to the annual gross domestic product (GDP), engages 44% of total labour force and contributes 60% to the national export. Pakistan has a total area of 80 Mha (million hectares) with 22 Mha arable land, out of which 17 Mha is under irrigation, mostly under canal irrigation. Due to the arid to semi-arid climate, the irrigation is predominantly necessary for successful crop husbandry in Pakistan.

The development of modern irrigation in Indo-Pakistan started in 1859 with the construction of the Upper Bari Doab Canal on Ravi River and with the passage of time the irrigation system of Pakistan grew up to the world's largest contiguous gravity flow irrigation system, known as the Indus Basin Irrigation System (IBIS). In the IBIS almost all irrigation canals are directly fed from rivers, while river flows carry heavy sediment loads. Irrigation canals receiving such flows get massive amounts of sediments, which are then deposited in the irrigation canals depending upon the hydrodynamic conditions of the canals. Sediment deposition in irrigation canals causes serious operation and maintenance problems. Studies reveal that silt reduces up to 40% of the available discharge in irrigation canals.

Researchers have been striving since long to manage this problem in a sustainable way and a number of approaches have been introduced in this connection. As a first step sediments are controlled at river intakes by silt excluders and ejectors. Then a canal design approach is adopted for keeping sediments in suspension and to distribute them as much as possible on the irrigated fields. Even then sediments tend to deposit in irrigation canals and become a serious problem in canal operation and maintenance, which then requires frequent desilting campaigns to keep water in the canals running. It causes a continuous burden on the national economy. In emerging and least developed countries, adequate and timely availability of funds for operation and maintenance is generally a problem. It causes delays in canal maintenance, which affects their hydraulic performance. Water is then delivered inadequately and inequitably to the water users.

The story becomes further complicated when it comes to downstream controlled demand based irrigation canals under flexible operation. In fixed supply based operation, canals always run at full supply discharge and such operation, generally, does not allow sediment deposition in the canal prism due to sufficient velocities. Whereas in demand based flexible operation the canals cannot run always at full supply discharge but instead the discharge is changing depending upon the crop water requirement in the canal command area. Such type of canal operation is not always favourable to sediment transport as under low discharges, flow velocities fall quite low and hence sediment deposition may occur in the canal prism. The questions arise here what sort of hydrodynamic relationships prevent sediment deposition in downstream controlled irrigation canals and how these relationships can be adopted, while catering crop water requirements of the command area? How the maintenance needs can be minimized by managing sediment transport through better canal operation?

This study has been designed to investigate such type of relationships and practices in order to manage sediment transport in downstream controlled demand based irrigation canals and to attain maximum hydraulic efficiency with minimum maintenance needs. The hypothesis of the study states that in demand based irrigation canals the volume of silt deposition can be minimized and even the sediments which deposit during low crop water requirement periods can be re-entrained during peak water requirement periods. In this way a balance can be maintained in sediment deposition and re-entrainment by adequate canal operation.

Two computer models have been used in this study, namely, Simulation of Irrigation Canals (SIC) and SEdiment TRansport in Irrigation Canals (SETRIC). Both models are one-dimensional and are capable of simulating steady and unsteady state flows (SETRIC only steady state flows) and non equilibrium sediment transport in irrigation canals. The SIC model has the capability to simulate sediment transport under unsteady flow conditions and can assess the effect of sediment deposition on hydraulic performance of irrigation canals. Whereas the SETRIC model has the advantage of taking into account the development of bed forms and their effect on resistance to flow, which is the critical factor in irrigation canal design and management. In the SETRIC model, a new module regarding sediment transport simulations in downstream controlled irrigation canals has been incorporated.

The study has been conducted on the Upper Swat Canal – Pehure High Level Canal (USC-PHLC) Irrigation System, which consists of three canals, Machai Branch Canal, PHLC and Maira Branch Canal. The Machai Branch Canal has upstream controlled supply based operation and the two other canals have downstream controlled demand based operation respectively. These canals are interconnected. The PHLC and Machai Branch canals feed Maira Branch Canal as well having their own irrigation systems. PHLC receives water from Tarbela Reservoir and Machai Branch Canal from the Swat River through USC. Water from Tarbela Reservoir, at present, is sediment free, whereas the water from Swat River is sediment laden. However, various studies have indicated that soon Tarbela Reservoir will be filled with sediments and will behave as run of the river system. Then PHLC will also receive sediment laden flows. The design discharges of Machai, PHLC and Maira Branch canals are 65, 28 and 27 m^3/s respectively. The command area of the USC-PHLC Irrigation System is 115,800 ha.

The USC-PHLC Irrigation System has been remodelled recently and water allowance has been increased from 0.34 l/s/h to 0.67 l/s/h. The upper USC system, from Machai Branch head to RD 242 (a control structure from where the downstream control system starts), was remodelled in 1995, whereas the system downstream of RD 242 was remodelled in 2003. The upper part of Machai Branch Canal up to an abscissa of about 74,000 m is under fixed supply based operation, whereas the lower part of Machai Branch Canal, Maira Branch Canal and the PHLC are under semi-demand based flexible operation. The semi-demand based system is operated according to crop water requirements and follows a Crop Based Irrigation Operations (CBIO) schedule. When the crop water demand falls below 80% of the full supply discharge, a rotation system is introduced among the secondary offtakes. During very low crop water requirement periods the supplies are not reduced beyond a minimum limit of 50% of the full supply discharge because of the canal operation rule.

The study consisted of fieldwork of two years in which daily canal operation data, monthly sediment inflow data in low sediment periods and weekly sediment data in peak concentration periods were collected. Three mass balance studies were conducted in which all the water and sediment inflows and outflows were measured with suspended sediment sampling at selected locations along the canal and boil sampling at the offtaking canals, immediately downstream of the head regulators. Further in the four months during the peak sediment season June, July, August and September, mass balance studies were conducted by boil sediment sampling in order to estimate water and sediment inflow to and outflow from the system. To determine the effect of sediment transport on the canals' morphology, five cross-sectional surveys were conducted and changes in bed levels were measured. On the basis of these field data the two computer models, used in this study, were calibrated and validated for flow and sediment transport simulations.

The downstream control component of the system is controlled automatically and the PHLC has been equipped with the Supervisory Control and Data Acquisition (SCADA) system at the headworks. Any discharge withdrawal or refusal by Water Users Associations (WUA) through offtaking secondary canals, or any discharge variation in the inflow from Machai Branch Canal is automatically adjusted by the SCADA system at Gandaf Outlet, the PHLC headworks. The SCADA system has Proportional Integral (PI) discharge controllers. The study found that the existing PI coefficients led to delay in discharge releases and resulted in a long time to achieve flow stability. The discharge releases showed an oscillatory behaviour which affected the functioning of hydro-mechanically operated downstream control "*Aval Orifice*" (AVIO) and "*Aval Surface*" (AVIS) gates. After calibration and validation of the model the PI controllers were fine-tuned and proposed for improved canal operation, which would help in system sustainability and in improved operational efficiency of the canals.

Field data show that during the study period sedimentation in the studied irrigation canals remained within control limits. The incoming sediment loads were, generally, lower than the sediment transport capacities of the studied irrigation canals. Hence this incoming sediment load was transported by the main canals and distributed to the offtaking canals. The sediment transport capacities of the studied irrigation canals were computed at steady and unsteady state conditions. The canal operation data showed that the system was operated on Supply Based Operation (SBO) approach rather than CBIO. The morphological data revealed that there was no significant deposition in the studied canals. Therefore there was no particular effect on the canal operation and the hydraulic efficiencies, attributed to sediment transport.

As mentioned earlier, the Tarbela Reservoir will soon be filled with sediments and consequently PHLC will get sediment laden flows from the reservoir. Various studies have been taken into account to project the time when sediment laden flows will flow into the PHLC and what will be the characteristics and concentrations of the incoming sediments to the PHLC from the reservoir. The studies project that the sediment inflow from the Tarbela Reservoir will be much higher than the sediment transport capacities of the PHLC and Maira Branch Canal under full supply discharge conditions. This scenario will create sediment transport problems in downstream controlled canals, particularly when they will be operated under CBIO.

Various management options have been simulated and are presented in order to better manage sediments in the studied canals under the scenario of sediment inflow from Tarbela Reservoir. The hydraulic performance of downstream controlled canals will be affected under this scenario and frequent maintenance and repair will be required to maintain the canals. Various options have been analysed to deal with the problem. The study presents a sediment management plan for downstream controlled irrigation canals by improvements in canal design and operation in combination with the need of settling ponds at the canal headworks.

Currently sedimentation in the irrigation canals under study is not a big issue for canal operation and maintenance (O&M). However, it would emerge as a major problem when sediment discharge from the Tarbela Reservoir starts. The canals' maintenance costs will soar and the hydrodynamic performance of these canals will also be affected. In this study, a number of ways have been evaluated and proposed to deal with the approaching problem of sediment transport in these irrigation canals in order to keep their hydraulic performance at desired levels and to minimize the maintenance costs. The first and the foremost effect of sediment deposition will be reduction in canals' flow conveyance capacities, which will result in raise of water levels. The raise of water levels will cause a reduction in water supply to the canals due to automatic flow releases. It can be dealt with by a temporary and limited raise in target water levels depending upon the maximum headloss at the downstream AVIS/AVIO cross regulator. Further, to minimize the effect of water level raise on discharge through the AVIS/AVIO gates, the decrement in such canals can be kept relatively small, in order to make the gates less sensitive to water level changes. Further, for efficient withdrawal of sediment to the secondary canals, it is needed to locate the secondary offtakes close to AVIS/AVIO cross regulators on the downstream side. More sediment will be discharged because the turbulent mixing of sediment at the downstream side of the control structures keeps more sediment in suspension. In addition, during the peak sediment concentration periods, the canals need to be operated at supply based operations, in order to minimize the deposition.

Sediment transport in general and in irrigation canals in particular, is one of the most studied and discussed topic in the field of fluid mechanics all over the world. It also has been studied extensively in Indus Basin in order to design and manage irrigation canals receiving sediment laden flows. The outcome of Lacey's regime theory and the subsequent work are the result of these studies. In addition to regime method various other methods like permissible velocity method, tractive force method and the rational methods, etc., have been developed for stable canal design. Anyhow, as a matter of fact, the management of sediment transport in irrigation canals is still a challenging task even after all these investigations and studies. Because most of the knowledge on sediment transport is empirical in nature, most sediment transport formulae have inbuilt randomness, which makes predictions difficult, when conditions are changed. It needs a lot of care while applying a sediment transport formula, developed under one set of conditions, to other situations. Therefore, it becomes extremely important to understand the origin of the development of the formulae and the limitations associated with them before applying some sediment transport formulae to different conditions and circumstances. The introduction of numerical modelling made it comparatively easy to test and shape the sediment transport relationships to some local conditions by running a variety of simulations and calibrating the formula in light of the field measurements. The sediment transport predictions can be made reliable in this way and can be used for further analysis.

1. Introduction and background

1.1   Background

Rivers originating from the Himalayan Region carry enormous amounts of sediment. Milliman and Syvitski (1992) stated that a third of the estimated contemporary global flux of about 18 Giga Tonnes per year (GT/yr) of fluvial suspended sediment is transported by the rivers of southern Asia. Of this total annual global flux, about 9% (1.6 GT/yr) is carried by five large rivers which originate in the Himalaya, the Brahmaputra, Ganga, Indus, Irrawaddy and Mekong. In general the sediment fluxes of these rivers are substantially higher than those of rivers draining other mountainous regions of the world. If dams and barrages had not been constructed, the three principal rivers of the Indian Sub-continent would annually deliver a combined 1.2 GT/yr (7% of the total global flux) to the coast (Milliman and Meade, 1983; Milliman and Syvitski, 1992). An earlier estimate by Meybeck (1976) suggested that the Brahmaputra, Ganga and Indus transport an aggregated pre-dam flux of 1.6 GT/yr.

Indus River is the major river of Indus Basin with its five main eastern tributaries the Ravi, Sutlej, Beas, Chenab and Jhelum rivers and one western tributary, the Kabul River. All of these streams carry enormous loads of sediments. For instance Sutlej River transports 13 MCM (million cubic metres) of sediments every year, Indus River carries a total sediment load of 166 MCM/yr at Tarbela, and estimates for Jhelum River are 26 MCM/yr (Ali, 1993). The water coming from different sub-basins of the Indus Basin carries about 435 MCM of suspended sediments every year. About 249 MCM of these sediments are deposited in reservoirs and irrigation canals (Water and Power Development Authority (WAPDA), 1985). Almost all of the irrigation canals of the Indus Basin are directly fed from these rivers and receive high amounts of silt.

The development of modern irrigation in the sub-continent started in 1859 with the completion of Upper Bari Doab Canal (UBDC) on Ravi River at Madhopur Headworks (now in India) and with the passage of time the irrigation system of Pakistan grew up to the world's largest contiguous gravity flow irrigation system, called the IBIS (Indus Basin Irrigation System). It encompasses the Indus River with its tributaries with three large reservoirs at Tarbela, Mangla, and Chashma and commands about 16 million ha of land. There are 23 barrages/headworks, 12 inter-river link canals and 45 canal commands extending for about 60,800 km to serve over 90,000 farmers' operated tertiary watercourses.

Irrigation canals receiving these river flows get massive amounts of sediments, for example the Upper Bari Doab Canal (UBDC) received almost 5,000 m^3 to 64,000 m^3 of sediments per day in a period of ten years, from 1939 to 1949 (American Society of Civil Engineers, 1972). Canals receiving such high amounts of silt load encounter, sometimes, severe maintenance problems. Like Marala-Ravi Link Canal faced severe sedimentation problems shortly after construction, which forced the authorities to construct a new headwork, causing heavy economic burden. As a result, the canal had lost one third of its original conveyance capacity. Jamarao Canal, one of the 43 main canal commands of Pakistan, receiving water from Indus River, has been raised up, since 1932, by 3 metres in its upper reaches because of sedimentation (Belaud and Baume,

2002). When this silt-laden water flows into an irrigation system, sediments tend to deposit in irrigation canals. If only a small proportion of sediment is deposited in irrigation canals, the subsequent changes in flow characteristics could result in major problems of operation and maintenance for the irrigation managers and farmers (Brabben, 1990). The sedimentation may reduce up to 40% of the available discharges in some irrigation canals. The reduction in conveyance capacity may cause problems in adequacy and proportionality of water distribution. Large budgets are frequently required for maintenance of such canals, which become a continuous burden on the economies.

Researchers have been striving since long to find practical solutions of such problems and a wide number of approaches have been developed to combat the situation. Effective sediment management and control starts at the river intakes where coarse sediments are excluded from the flow entering into the canal by silt excluders and then ejected further when entered into the canal by silt ejectors. Then by canal design approach the sediments are kept in suspension and distributed to the irrigated fields where they become part of soil and enhance the fertility of the land. Canal design approaches are further strengthened by adopting suitable flow regulation schedules such that the flow velocities in the canal may not drop so low that they allow a sedimentation process.

1.2   Rationale of the study

Traditional water allowance in Pakistan is 0.2-0.3 l/s/ha (litres per second per hectare) which is not sufficient to meet peak crop water demands. In the irrigation system under study the water allowance has been increased almost two-folds, from 0.3 l/s/ha to 0.7 l/s/ha, to meet the peak crop water demands. As crop water requirements do not remain same throughout the cropping season the canal supplies are adjusted accordingly. To keep a balance between supply and demand a new canal operational strategy was adopted in this irrigation system, called Crop Based Irrigation Operations (CBIO). Under these operations, flow conditions in the canals keep on varying throughout the cropping season, which has serious implications on the sedimentation process in the canals.

As mentioned earlier sedimentation in irrigation canals causes numerous operation and maintenance problems like reduction in conveyance capacity, inadequacy and inequity in water distribution and increased risk of canal breach due to reduction in freeboard, etc. Then huge amounts of money are required to maintain the canals, where adequate and timely availability of funds for operation and maintenance is generally a problem in emerging and least developed nations, which delays the maintenance of irrigation canals substantially.

Provincial Irrigation Departments are the custodians of the IBIS canals in Pakistan and every year they allocate funds for maintenance and repair (M&R) of the canals. A fixed rate or "yardstick" is utilized by the departments to determine the actual amount of the annual grant allocation to each unit for this purpose from the total amount of the annual budget approved by the provincial government for the department. The yardstick for repair and maintenance of large and small irrigation canals has been specified as US$ 677/km and US$ 253/km for main canals and distributaries respectively (Punjab

Irrigation and Power Department, 2008). Though, the amount of M&R funding seems reasonable, but usually less funds are released for he M&R than the required and consequently maintenance activities are carried out only on selected portions of the canals or concentrated on very few canals in any particular division or sub-division of the irrigation department. Under such pressing conditions of high sediment load to irrigation canals and limited availability of M&R funds, it becomes a quite challenging task to keep the canal's hydraulic performance at desired levels.

The situation becomes more complicated when it comes to downstream controlled automatic irrigation canals, where the hydrodynamic conditions in the canals are not usually allowed to be compatible with the needs of sediment transport, as the priority is shifted to meet the crop water demands instead of sediment control. Here a trade-off is required to operate canals at such conditions that it allows minimum sediment deposition while catering the crop water requirements.

1.3   Scope of the study

In this study the sediment related problems in irrigation canals have been analyzed with the standpoint of managing sediments by improving canal design and operation. The study has explained how sedimentation can affect automatically downstream controlled irrigation canals' hydrodynamic behaviour and its subsequent effects on the hydraulic efficiency of such canals. The relationships between canal operation and sediment transport have been presented and based on these improvements in downstream controlled canal design and operation for sediment control have been suggested. Some canal operation techniques have been developed and tested to minimize dredging/desilting costs in order to maintain the desired hydraulic performance of the canals. Additionally, the future problems resulting from increasing sediment inflow rates have also been highlighted along with their remedial measures. Mathematical models like SIC and SETRIC have been calibrated/developed, validated and have been used to simulate the hydrodynamic and sediment behaviour in the canals.

1.4   Research questions

The behaviour of sediment transport in downstream controlled irrigation canals is quite complicated due to ever changing flow conditions in the canals. In typical fixed supply based operation, canals always run at full supply discharge and such operation, generally, does not allow sediment deposition in the canal prism due to sufficient high velocities. Whereas in demand based flexible operation the canals do not run always at full supply discharge but instead the discharge keeps on changing, depending upon the crop water requirement in the canal command area. Such type of canal operation is not always favourable to sediment transport as under low discharges, flow velocities fall quite low and hence sediment deposition occurs in the canal prism. The research questions arise here that what sort of hydrodynamic relationships may prevent sediment deposition in downstream controlled irrigation canals and how these relationships can be adopted, while catering crop water requirements of the command area? How can the maintenance needs be minimized by managing sediment transport through better canal operation and is it practically possible that the deposited sediments during low flow conditions can be scoured again at high conditions?

1.5   Research hypothesis

The approaches of silt exclusion and ejection, and design of silt free irrigation canals have been proved fruitful but not capable enough to manage sediments in the flows entering into irrigation canals at a desired level, which lead to sediment deposition in irrigation canals. The concept of the study postulates that the sediment transport can be managed by improvement in canal operation. For example the amount of sediment deposited during low flow conditions can be scoured during high flow in demand based irrigation canals and thus minimize the maintenance needs. The other postulation states that by giving priority to sensitive canal reaches during dredging/desiltation campaigns the hydraulic efficiency of the irrigation canals can be improved by minimum maintenance.

1.6   Objectives

Improving hydraulic performance, managing sediment transport with improvements in canal operation and minimizing maintenance needs of the irrigation canals are the basic objectives of this study. To meet them, the detailed objectives of the study have been formulated as to:

- understand canals' hydrodynamic behaviour under supply and demand based operation;

- assess irrigation canals' hydraulic performance under existing sediment transport and with increased sediment load from the Tarbela Reservoir;

- develop an optimal canal operation plan to inhibit sediment deposition during peak concentration periods and to allow scouring of the deposited sediments, where possible, during low concentration periods, while catering the crop water requirements of the command area; and

- propose sediment management options for achieving maximum hydraulic efficiency with minimum desiltation requirements.

1.7   Introduction of the thesis

Chapters 1, 2 and 3 in this thesis describe in detail the nature of sediment transport problems in irrigation canals. Chapter 1 provides some background, need and objectives of the study. Chapter 2 sheds light on water resources and general socio-economic conditions of Pakistan. Chapter 3 is a review of past research in this field and also gives overview of the hydraulics of downstream control and flow automation in irrigation canals.

Chapter 4 is on sedimentation in Tarbela Reservoir, which is a major reservoir in IBIS and also is the source of water for the studied irrigation canals.

Chapter 5 presents modelling results of hydrodynamic behaviour of the system. Chapter 6 presents the results of field measurements on water and sediment flow, morphological changes in the canal and the hydrodynamic performance of studied irrigation canals.

Chapters 7 presents the modelling results of sediment transport and sediment management options in the downstream controlled irrigation canals.

Chapter 8 is about the development of the downstream control component in the SETRIC Model and its application to the irrigation canals under study.

The sediment management options in downstream controlled irrigation canals have been given in Chapter 9.

Chapter 10 is about the evaluation of the findings of this study.

2. Indus Basin Irrigation System in Pakistan

2.1 General information about Pakistan

Pakistan's total land area is 796,100 km^2. As shown in Figure 2.1, administratively it is divided into four provinces: Khyber Pakhtunkhwa (KPk) with 15.9% of the population, Punjab (55.1%), Sindh (22.9%) and Balochistan (4.9%), and several areas with special status, including the Federally Administered Tribal Area (FATA), the Northern Area (FANA), State of Azad Jammu and Kashmir, and the Capital Area of Islamabad. About half the land area comprises mountainous terrain, narrow valleys and foothills and other areas of limited productivity. Most of the remaining area lies in the Indus Plain, covering some 20 Mha of generally fertile alluvial land. The Indus Basin Irrigation System commands about 15 Mha and is the agricultural and economic centre of the country. All the major hydropower stations are also located in the Indus Basin.

Figure 2.1. Map of Pakistan with administrative boundaries (Source: Ministry of Water and Power, Pakistan)

2.2   Socio-economic settings of the country

It is estimated that Pakistan's population has reached up to 160 million and a major portion of it, about 67% lives in rural areas. About 32% of the people are living below the poverty line (Federal Bureau of Statistics, 1999). The incidence of poverty is higher among rural people as in rural areas it is 36.3% whereas in urban areas it is 22.6%. Poverty is high among non-farm rural people, small farmers, people having inequitable access to water and in the areas with saline groundwater as compared with areas that have sweet groundwater (World Bank, 2007).

Agriculture is the major activity in the rural areas and vital for food security and the economy of Pakistan. Out of the total land of 80 Mha of Pakistan, 21 Mha is under cultivation, annually producing about 32 MT (million tonnes) and 57 MT of food crops and cash crops respectively (Agricultural Statistics of Pakistan, 2006-2007). Out of the total industrial production of Rs. 1,150 billion (US$ 19 billion), 60% contribution is from agriculture dependent industries. Along with crops livestock is also an important sector contributing to the agricultural production. There are 143 million livestock heads in Pakistan (Agricultural Statistics of Pakistan, 2006-2007), which also contribute equally in the national economy. There were Rs. 29 billion leather and dairy products during the year 2006-2007.

Water is predominantly necessary for successful crop husbandry due to the arid to semi-arid climate of Pakistan. Owing to agricultural water needs Pakistan has world's largest contiguous gravity flow irrigation system with 19 barrages, 45 canal systems with a command area of 17 Mha, with annual diversion of 139 BCM (billion cubic metres) (Halcrow, 2003). Further, there are about 0.95 million tubewells in Pakistan, extracting an amount of 62 BCM of water. Total water availability in Pakistan is 184 BCM/yr with 179 BCM/yr in Indus Basin river system.

2.3   Water resources

The Indus River and its tributaries (Jhelum, Chenab, Ravi, Sutlej and Beas rivers on the East and Kabul River on the West), which drain an area of 945,000 km^2, are the main sources of surface water in Pakistan, which bring an average of 187 BCM of water annually. This includes 174 BCM/yr from the western rivers (Indus, Kabul, Chenab and Jhelum) and 10 BCM/yr from the eastern rivers (Ravi and Sutlej). In addition to the Indus Basin there are several smaller basins in Balochistan, namely the Kharan Desert Basin, which is a closed basin, and the Makran Coastal Basin, which discharges to the sea. The rivers in the Makran Basin have an average flow of about 4 BCM/yr, and the rivers in the Kharan Basin have an average flow of about 1 BCM/yr.

The Indus Basin is mainly alluvial and is underlain by an unconfined aquifer covering about 6 Mha in surface area. The main sources of recharge are direct rainfall and infiltration through the alluvium from rivers, the irrigation system and from the irrigated fields and the annual total recharge is 56 BCM/yr. From the perspective of groundwater use, it is estimated that about 51 BCM of groundwater is abstracted for irrigation use and for domestic water supplies (Ministry of Water and Power (MoWP), 2002).

2.4   Indus Basin Irrigation System

Surface water is the major source of irrigation water in Pakistan, which is diverted from rivers to irrigation canals through barrages and headworks. The canal irrigation system covers a gross area of 16 million ha of which 88% is cultivable. There are 48 principle canals, emerging out of 20 river diversion structures. The cumulative operating capacity of these canals is 7,320 m^3/s and their annual conveyance capacity is 331 BCM. These canals traverse about 61,000 km to command the 15.5 million ha of culturable area through 90,000 watercourses. Each watercourse serves about 160 ha of land on average. In addition, there are 23 barrages, 45 main canals, 12 inter-river link canals transferring bulk water supplies from the western rivers to the eastern rivers.

Pakistan has two major reservoirs for irrigation as well as for hydropower; Mangla Reservoir with gross water storage capacity 7 BCM and Tarbela Reservoir with gross storage capacity 14 BCM. The Chashma Barrage has also a big reservoir which helps in the irrigation of millions of hectares of agricultural lands. Further, there are about 0.95 million tubewells in Pakistan, extracting an amount of 62 BCM/yr of water (Agriculture Statistics of Pakistan, 2006-2007).

2.5   History of irrigation in the Indus Basin

Since the beginning of the civilization the main source of subsistence of the people in Indus Basin is agriculture. So irrigation remained in focus since those times. The evidence of irrigation can be found during the per-historic period, during the Harrapan Civilization 2300 BC to 1500 BC and then 1500 BC to 1526 (Fahlbusch et al., 2004). Anyhow, the modern engineering based irrigation system development started with the British entry into the Indus Basin in early nineteenth century, when British rule was extended to the tracts between the Yamuna and Sutlej rivers in 1819. The development of IBIS can be divided into three major phases, the British rule up to 1947, 1947 to Indus Waters Treaty in 1960 and after 1960. However, the major developments took place during the British rule.

In the beginning, the old inundation canals were developed and improved. The Punjab Public Works Department (PWD) manned by army engineers was organized in the middle of the nineteenth century and developed inundation canals till the last decade of the nineteenth century (Ali, 1963). Various inundation canals were developed on all the rivers in Punjab, Khyber Pakhtunkhwa and Sindh like Hasli Canal, later named as Upper Bari Doab Canal (UBDC) from Ravi River, Kabul Canal in North-western Frontier Province (now Khyber Pakhtunkhwa), Desert Canal, Begari Canal, Sukkur Canal, Garh Canal, Nara and Jamrao Canal in Sindh.

The systematic development of irrigation canals with weir controlled supplies started in 1850 and first such canal built, was the UBDC. The headworks were constructed at the Ravi River at Madhopur. The next such large project was the development of Sirhind Canal from the Sutlej River at Ropar to irrigate the districts of Ludhiana, Ferozpur and Hissar, etc. Then a network of canals was developed all over the Indus Basin like Sidhnai Canal (1883) taking off from a straight reach of Ravi River between Tulamba and Sarai Sidhu, Lower Chenab Canal (1890) from Khanki Headworks, Lower Jhelum Canal (1901) from the left bank of Jhelum River at Rasul.

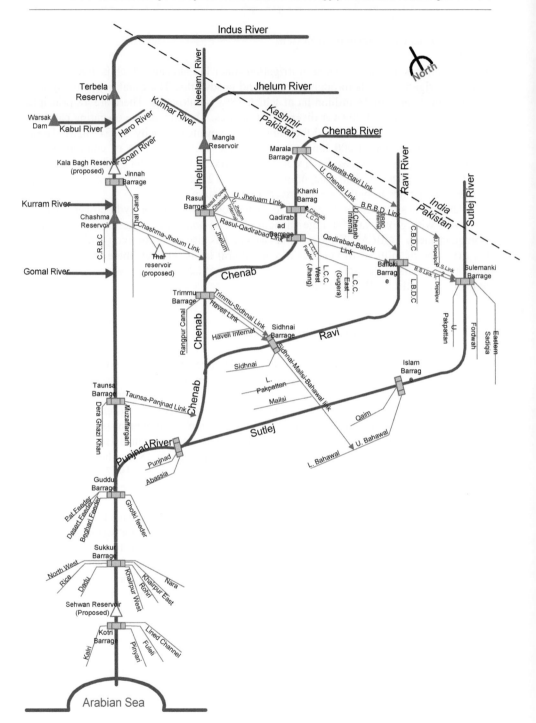

Figure 2.2. Schematic map of the Indus Basin Irrigation System

The Triple Canal Project (Upper Jhelum Canal, Upper Chenab Canal and Lower Bari Doab Canal) was envisaged in 1905. This was a system of link canals, as well as having their own system for irrigation, starting from Jhelum River, through Upper Jhelum Canal, to Chenab River, and then to Ravi River through Upper Chenab Canal. The project was completed in 1915. The gigantic Sutlej Valley Project (1921) was designed to replace the old-shutter type weirs with gate-controlled Barrages. Four weirs at Ferozpur, Sulemanki, Islam and Panjnad were constructed. Four canals, the Pakpattan, Dipalpur, Eastern and Mailsi canals were constructed in 1933 under this project. Then Haveli Canal and Rangpur Canal were completed in 1939 from the Trimmu Headworks. Silt excluders and ejectors were used for the first time in this project. In Sukkur Barrage Project (1932), the Sukkur Barrage along with seven canals capable of carrying 1,350 m^3/s of water was constructed. Rice Canal, Dadu Canal, Eastern Nara Canal, Rohri Canal, Khairpur West Feeder were the major canals. Kotri Barrage and Guddu Barrage were also planned, but constructed after the partition from India. Overall around 30 out of 43 canal commands in Pakistan were built or improved upon by 1947. The remaining 13 were improved upon during the years 1955 to 1962. These works made it possible to irrigate an area of 10.5 Mha, that was larger than the total irrigated area in any country of the world.

In August 1947 Pakistan got independence from the British rule and became an independent state. After partition, the momentum generated for irrigation development before independence, continued with the same vigour. The Warsak Dam and Taunsa Barrage were constructed. The Abbasia Canal was extended and Thal Canal Project was undertaken. After partition, Pakistan constructed some link canals in order to meet any possible shortage in eastern rivers from the western rivers. The Bombanwala-Ravi-Badian-Dipalpur Link, Balloki-Sulemanki Link and Mara-Ravi Link were constructed in this regard. In Sindh, the Kotri Barrage was completed in 1955. There were three feeder canals on the left side the Pinyari, Phuleli and Akram Wah Canal. On the right side, the Kalri Bagh Feeder started functioning in 1958.

The partition divided the Indus Basin as well, with Pakistan remaining the lower riparian. The two major headworks, one at Madhopur at Ravi River and the other at Ferozpur on Sutlej River, on which the irrigation supplies in Punjab were dependent, now fell in the Indian Territory giving rise to administrative problems regarding the regulation and supply of water through existing facilities. Various conflicts arose between the two countries on water distribution of the Indus Basin Rivers. Ultimately, with the help of the World Bank (the then International Bank for Reconstruction and Development) a treaty was signed between the two countries in 1960, the famous Indus Waters Treaty. According to the treaty, India got exclusive rights on the three eastern rivers, the Sutlej, Beas and Ravi and Pakistan got the same rights for the three western rivers, the Indus, Jhelum and Chenab. In order to meet the shortages in the irrigation canals, dependent upon eastern rivers, some mega projects were undertaken by construction of two large dams, the Mangla Dam (1966) and the Tarbela Dam (1976), construction of eight large capacity link canals, six barrages and remodelling of three of the existing inter-river link canals. There was no big irrigation canal project executed after these developments. However, the Government of Pakistan approved the construction of three new irrigation canals in 2002, the Raini Canal, the Greater Thal Canal and Kachhi Canal in IBIS. Amongst these the former two are under construction.

3. Sediment transport and flow control in irrigation canals

3.1. General information on sediment transport and modelling

Sediment transport starts when shear forces applied by the flow overcome the weight of the particle, it is the initiation of motion. Then depending upon the hydrodynamic conditions and sediment characteristics, particles move in the form of bed load or suspended load. The work done by various researches on these aspects of sediment transport has been summarized below.

3.1.1   Incipient motion

The history of description of incipient motion (initiation of motion) starts with the Kramer's (1935) definition of weak, medium and general movement of sediment particles. Later Shields found that the motion of a particle is controlled by two opposing forces: the applied forces and the resisting forces. A particle will move if the applied forces, from flow hydrodynamics, overcome the resisting forces, weight of the particle (Shields, 1936). The threshold point for movement, or critical condition, occurs when the forces of flow are exactly balanced by the submerged weight of the particle. Shields did not fit a curve to the data but indicated the relationship between τ_{*c} (critical non-dimensional shear stress) and R_{*c} (critical boundary Reynolds number) by a band of appreciable width. The Shields' curve was first proposed by Rouse (1939).

Shields' results were widely accepted although some researchers have reported somewhat different values for the parameters like: Egiazaroff (1950), Tison (1953), Lane (1955). All of these researchers obtained curves having the general shape, like Shields' curve. Vanoni (1966) preferred Shields' diagram to other criteria to determine the initiation of motion. However, Barr and Herbertson (1966) expressed their dissatisfaction with the idea of critical shear stress as the criterion for initiation of motion. Yang (1973) also criticized Shields' diagram as the best criterion for incipient motion because of the following reasons:

- selection of shear stress instead of average velocity;
- shear stress is a measure of the intensity of turbulent fluctuations (Shames, 1962), and the present knowledge of turbulence is limited;
- Shields' criterion uses the concept of laminar sublayer and the laminar sublayer should not have any effect on the velocity distribution when the shear velocity Reynolds number is greater that 70;
- Shields simplified the problem by neglecting the lift force and considers the tangential force only;
- because the rate of sediment transport cannot be uniquely determined by shear stress, it is questionable whether critical shear stress should be used as the criterion for incipient motion for the study of sediment transport (Brooks, 1955 and Yang, 1972).

The original version of Shields' diagram does not apply adequately to sediment having low specific gravity and a small diameter. Mantz (1977) extended the range of Shields' diagram to smaller particles to overcome the difficulty with small particles. Zanke (2003) found on the basis of the analytical formulation that the initiation of sediment motion might be described solely by the angle of repose of the grains and turbulence parameters.

3.1.2   Bed load sediment transport

Bed load is the mode of transport of sediments where the particles glide, role or jump but stay very close to the bed (Graf, 2003). The bed load generally consists of coarser particles. It is very important in sediment transport as it controls the shape, stability, and hydraulic characteristics of the channel. The bed load equations are generally grouped into three following types:

- *DuBoys-type equations* that utilize a shear stress relationship;
- *Schoklitsch-type equations* that utilize a discharge relationship;
- *Einstein-type equations* that are grounded in statistical considerations of lift forces.

DuBoys (1879) assumed that the bed material moves in layers of thickness d, and that mean velocity of the successive layers increases linearly towards the bed surface. Schoklitsch (1930) stated that the average bed shear stress in DuBoys equation is a poor criterion when applied to field computations, because the shear distribution in the channel cross-section is quite non-uniform. Einstein (1942) developed a relationship for sediment transport based on the probability theory. Ignoring drag force effects on a particle. Einstein assumed that a particle would be dislodged from the channel bed if the lift force exceeds the submerged weight during an exchange time. The particle would move a distance of λd, where d is the particle diameter and λ is the general constant, depending on the probability of erosion, before striking the surface again. If the forces at the point of impact are such that lift exceeds the submerged weight, the particle will take another hop of distance λd before striking the surface again. This continues until the particle comes to a point where the lift is less than the submerged weight. Simons and Senturk (1992) indicated that bed load amounts to about 5-25% of the suspended load. They emphasized that although this amount is a relatively small proportion, it controls the shape, stability, and hydraulic characteristics of a channel.

Numerous bed load discharge equations have been derived, but only a limited number of field studies are available for validation or for the further development of formulae. Major reviews of the bed load discharge formulae were performed using either laboratory data or field data (Vanoni et al., 1961, Shulits and Hill 1968, White et al., 1973, American Society of Civil Engineering (ASCE) Task Committee, 1975, Carson and Griffith, 1987, Gomez and Church, 1989, Reid et al., 1996, Lopes et al., 2001 and Zanke, 2003).

Harbersack and Laronne (2002) concluded that a formula requires selection according to applied conditions. When partial transport conditions occur, equilibrium-based formulae are not applicable. Because there is a gradual increase in the mobility of certain grain sizes with an increase in shear stress until near-equal mobility is attained.

This was also found by Gomez and Church (1989) that the Meyer-Peter and Muller (1948) equation fails to predict motion for a significant proportion of the data. This was also noted by laboratory and field measurements by Zanke (1990). Gomez and Church (1989) suggested the use of stream power equations for estimation of the magnitude of transport with limited hydraulic information. The unmodified shear stress formulae such as the one of Parker (1990) under predict the bed load flux (Reid and Laronne, 1995).

3.1.3   Suspended sediment transport

In suspended mode of sediment transport the sediment particles displace themselves by making large jumps, but remain (occasionally) in contact with the bed load and also with the bed (Graf, 2003). The suspended load usually consists of finer particles, such as silt and clay. There are two states of suspended sediment transport, equilibrium condition, no deposition no scouring, and non-equilibrium condition when either of the phenomena can take place.

Rubey (1933) developed the basic relationship between the work, energy expenditure of a stream and the quantity of the sediment transported by the stream. The approach of Rubey was further studied by Knapp (1938), and similar ideas were put forth independently but later by Bagnold (1956, 1966) in terms of stream power. For determining the suspended sediment transport in *equilibrium conditions* the Exchange theory is commonly used (Rous, 1937). Einstein (1950) developed a relationship assuming that the bed layer has a thickness equal to two times the particle diameter and there is no suspension within this layer. He developed the suspended sediment discharge equation for each particle size.

For non-equilibrium conditions the sediment suspended motion in the fluid caused by the turbulence effects is assumed to be analogous to the diffusion-dispersion of mass in open channel flows and can be derived considering the conservation of mass which states that the change rate of dispersant weight in the control volume is equal to the sum of the change rates of weight due to convection and diffusion. Bagnold (1966) developed a simple relationship by equating sediment transport rate to work rate. His equation is based on the assumption that mean velocity of fluid and suspended sediment is the same.

The theoretical approaches used to estimate suspended sediment discharge in streams are mainly the energy approach and the diffusion-dispersion approach. The diffusion-dispersion theory is recommended over the energy approach because experimental evidence indicated that it better fits to observed data (Simons and Senturk, 1977). Galappatti (1983) developed a depth-integrated model for suspended sediment transport in unsteady and non-uniform flow.

3.1.4   Total sediment load

Total load is the summation of the bed load and suspended load. A large number of relationships have been developed for total load prediction in the flow. Some widely used predictors have been discussed in the following:

- *Bagnold (1966).* Bagnold's formula is based on the energy concept. He assumed that the power necessary to transport the sediments is proportional to the loss of potential energy of the flow per unit time, and he called it "available power";

- *Engelund and Hansen (1967)*. They established the relationship between transport and mobility parameters for predicting sediment transport capacity under equilibrium conditions, which is also based on the energy concept;
- *Ackers and White (1973)*. This method describes sediment transport in terms of dimensionless parameters: D* (grain size sediment parameter, F_{gr} (mobility parameter) and G_{gr} (transport parameter);
- *Brownlie (1981)*. This method is based on a dimensional analysis and calibration of a wide range of field and laboratory data, where uniform conditions prevail. The sediment discharge in ppm (parts per million) is calculated by using a relationship among C_f (Coefficient of transport rate, 1.00 for laboratory conditions and 1.27 for field conditions, F_g (grain Froude number), F_{gcr} (critical grain Froude number), S (bed slope), R (hydraulic radius) and d_{50} (median diameter);
- *Yang (1984)* He used the basic form of the unit stream power equation and developed the relationship for gravel transport. He used the dimensional analysis and multiple regression methods to derive equations for computing discharge in sand bed streams;
- *Van Rijn Method (1984a and 1984b)*. The total sediment load transport by the Van Rijn method can be computed by the summation of the bed and suspended load transport. The Van Rijn method presents the computation of the bed load transport q_b as the product of the saltation height, the particle velocity and the bed load concentration. It is assumed that the motion of the bed particles is dominated by gravity forces;
- *Karim-Kennedy (1990)*. They obtained the total sediment discharge formula from nonlinear regression using a database of 339 river flows and 608 flume studies. They used some dimensionless ratios with a calibration data set and performed nonlinear regression analysis to develop dimensionless sediment discharge relationships.

3.2.   Evaluation of sediment transport formulae

Meadowcroft (1988) did a comparison of six sediment transport formulae (Ackers and White, Brownlie, Engelund and Hansen, Einstein, Van Rijn and Yang) against data collected from field sites restricted to the range of discharges, sediment sizes and concentrations commonly encountered in irrigation canals. The main characteristics, of the data used, were discharge between 1 to 200 m³/s, d_{50}, the median particle size, between 0.1 and 1 mm, concentration less than 7,100 ppm, Froude number between 0.1 and 0.99 and bottom shear stress between 0.65 and 10 N/m². The author concluded that the Engelund and Hansen sediment transport equation could be used in the absence of any local data. It is by a small margin the most accurate method and is the simplest one to use.

Woo (1988a) used a parameter sensitivity analysis and did some numerical experiments in a theoretical channel to assess the Einstein, Ackers and White, Yang, Toffaletti (1969), Engelund and Hansen (1967), Colby, Shen and Hung (1971) and Van Rijn (1986) formulae. The main conclusions were that the Einstein's, Toffaletti's and

Colby's formulae were very sensitive to the flow velocity, while the Engelund and Hansen and Yang formula were the least sensitive to that parameter. The Ackers and White formula was too sensitive to the particle size and it should not be used for fine sediments. The formula of Van Rijn (1986) and of Engelund and Hansen (1967) predict larger sediment transport than the other formulae, while Einstein and Toffaletti (1969) formulae predict smaller sediment transport.

Nakato (1990) tested eleven sediment transport formulae against field data of the Sacramento River. The bed material size was classified ranging from fine sand to coarse sand. The formulae of Ackers and White, Einstein and Brown, Einstein and Fredsoe, Engelund and Hansen, Inglis and Lacey, Karim, Meyer-Peter and Mueller, Van Rijn, Schoklitsch, Toffaletti, and Yang were tested. To evaluate the formulae for suspended transport, only suspended material larger than 0.62 mm was used. The Toffaletti formula proved to be the best among all formulae tested. Yang's predictions seemed to be very close to the measured suspended load discharge at higher range of sediment discharge.

Lawrence (1990) compared the performance of four bed friction and six sediment transport predictors with field data. The data set was restricted to the range of discharges, and sediment sizes and concentration commonly encountered in irrigation canals. He concluded that for general use the combination of the Van Rijn friction predictor with the Engelund and Hansen transport equation provides the best predictions.

Worapansopak (1992) evaluated five sediment transport formulae for the total suspended sediment. These were the formulae of Bagnold, Bruk, Celik, modified Ceilk, and Vligter. He used the Nakato (1990) data set to evaluate the suspended sediment transport formulae. Fine sediment and flume data were also used for this evaluation. The main conclusion was that all the formulae overestimate the sediment transport. In some cases the formulae lead to unacceptable results. The modified Celik transport formula was recommended to be used in irrigation canals, although the author recommended more research to verify the modified formula.

3.3.   Sediment transport modelling

Sediment transport modelling has been divided into four classes according to the number of space dimensions and spatial orientation; the (quasi-) three-dimensional (3D), two-dimensional vertical (2DV), two-dimensional horizontal (2DH) and one dimensional (1D) (Paudel, 2002).

3.3.1. One dimensional (1D) models

One dimensional (1D) models are mostly in use for simulating long-term morphological changes in irrigation canals. For example a popular models like SOBEK (Deltares, 2009) is a 1D model which can simulate steady and unsteady state flow in rivers and irrigation canals including simulations of other physical parameters like salt intrusion river morphology and water quality. Similarly MIKE 11 (Danish Hydraulic Institute, 2008) is a 1-D hydrodynamic model which permits computation of the sediment transport in rivers. There are many other models like HECRAS (United States (US) Army Corps of Engineers, 2010), Sediment and River Hydraulics-One Dimension (SRH-1D) Model (US Bureau of Reclamation, 2010), SEDICOUP (Belleudy and SOGREAH, 2000),

FLUVIAL-12 (Howard, 1998; 2006), HEC2SR (Simons et al., 1984) and IALLUVIAL (Karim and Kennedy, 1982) that have been developed for flow and sediment transport modelling in open channels.

Shou-Shan (1989) reviewed twelve computer sedimentation models developed and implemented in the United States of America; HEC2SR (Simons et al., 1984), HECG, TABS2, IALLUVIAL (Karim and Kennedy, 1982), STARS, GSTARS, CHARACTERISTICS, CHARIMA, SEDICOUP, FLUVIAL 12, TWODER and RESSED models. His main conclusions were:

- computer modelling of sedimentation problems was still in the development stage and exact representation is not possible;
- models had limited capabilities of modelling the effect of channel geometry and morphological changes;
- all models produced significantly different results even when they were run with the same set of input.

Bhutta et al. (1996) used a computer based hydraulic model, RAJBAH, to assess the utility of hydraulic models to assist and support canal system managers in planning and targeting maintenance activities on secondary canals. Their study confirmed that the suitably calibrated hydraulic simulation models can be effectively used in a decision support planning capacity to target and prioritize maintenance inputs for secondary canals in the irrigation systems of Pakistan's Punjab.

3.3.2. Two dimensional (2D) models

Two *dimensional vertical models (2DV)* are usually applied to predict transport rates, sedimentation and erosion in rivers, estuaries and coastal waters.

The two *dimensional horizontal models (2DH)* are based on the depth-integrated equations of motion in combination with the depth integrated sediment transport model as suggested by Galappatti and Vreugdenhil (1985).

FLUVIAL-12 (Chang and Hill, 1976), HEC6 (Thomas and Prashum, 1977), HEC2SR (Simons et al., 1984), IALLUVIAL (Holly et al., 1985), and ONED3X (Lai, 1986), stayed MOBED 2 (Spasojevic and Holly, 1988), Delft3D (Deltares, 2010) are two-dimensional sediment transport models which, are mostly used.

Guo et al. (2002) presented a two-dimensional horizontal (2DH) mathematical model for non-uniform suspended sediment transport to simulate riverbed deformation. Verification with laboratory data showed that the model had a good ability to simulate channel bed variation. Lee and Hsieh (2003) also developed a tow dimensional mathematical model which is capable of simulating scouring and deposition behaviour in a channel network. The model was applied to non-equilibrium sediment transport and unsteady flow conditions. The application of this model to the Tanhsui River System in Taiwan gave convincing results.

3.3.3. Three dimensional (3D) models

In the three dimensional models (3D) basically two approaches are used. First the depth integrated approach, introduced by Galappatti and Vreugdenhil (1985) and the second one is a three dimensional approach as applied by Sheng and Butler (1982), Muller (1983), Wang and Adeff (1986), O'Conner and Nicholson (1988), Van Rijn and Meijer (1988). Application of the depth integrated approach is limited to situations where the difference between the local actual suspended sediment transport and local equilibrium transport is relatively small. Three-dimensional models are becoming more popular but are only applied to predict the initial rate of sedimentation or erosion. For long-term morphological changes these models have limited use. Delft3D (Deltares, 2010) is a three dimensional model which simulates time and space variations of six phenomena and their interconnections. It consists of an advanced integrated and well-validated modelling environment for six modules that are linked to one-another, which are hydrodynamics [Delft3D-FLOW], waves [Delft3D-WAVE], water quality [Delft3D-WAQ], morphology [Delft3D-MOR], sediment transport [Delft3D-SED], and ecology [Delft3D-ECO].

Application of the 1-D, 2-D or 3-D models depends upon the nature of the problem and the area. Generally the 2-D and 3-D models are applied for studying the sediment transport problems in lakes, reservoirs, rivers, estuaries and oceans. The 1-D models are generally applied in irrigation canals and also in river sedimentation studies, where they give desirable results.

3.4.   Sediment transport modelling in irrigation canals

Sediment transport in irrigation canals differs in a number of ways from the natural channels and rivers. The sediment transport formulae for open channels have been developed generally for natural rivers. Though irrigation canals and rivers are similar in many ways, there are also many differences like the width to depth ratios, the control structures, sediment sizes, etc. In irrigation canals the objective of sediment transport calculations is usually to make predictions of morphological changes. The time frame of calculation is at least one irrigation season and may extend to several seasons. Hence the interest here would have to be to know what happens to the canal in the long run rather than within few hours or days. The level of sophistication of the mathematical model used must also be decided upon this context (Galappatti, 1983). Rivers are natural channels whereas canals are man-made. In rivers the effect of side banks on the flow resistance and on sediment transport are generally neglected because of the large width to depth ratios but in irrigation canals, the effect of side walls cannot be neglected (Mendez, 1998). He further stated that the river models cannot be successfully applied to irrigation canals. Keeping in view these differences he developed a 1D mathematical model SEdiment TRansport in Irrigation Canals (SETRIC) for sediment transport simulations particularly in irrigation canals. He introduced the concept of composite hydraulic roughness in this model. He took into account the effect of varying water depths on the side slopes and the boundary conditions imposed by the side slopes on the flow and velocity distribution.

Paudel (2002, 2010) made some significant improvements in the SETRIC model. He made a new input interface for the model. He made certain modifications for choosing roughness options. In the beginning only Nikuradse's roughness parameters could be entered but he added the option for Chezy's and Manning's roughness coefficients. He incorporated the effect on bed raise, because of deposition, on sediment conveyance at the structures. He improved the model to simulate the changing sediment concentration and size at different periods. Then he applied successfully this model to the Sunsary Morang Irrigation Scheme in Nepal for sediment transport predictions. He also compared the hydraulic calculations with other models and found satisfactory results.

Timilsina (2005) developed a Convection Diffusion Model in Sediment Transport in Irrigation Canals (CDSETRIC) for sediment transport simulations under non-equilibrium conditions, which differs from SETRIC in the computational procedure of non-equilibrium sediment transport. He used the computational approach suggested by Lee et al. (1997) instead of Galappatti's depth integrated model, for simulating deposition patterns of the suspended sediment in non-equilibrium processes. This method includes a sediment continuity equation, a sediment concentration convection-diffusion equation and a bed load equation. Lee et al. (1997) separated the convective diffusion equation into three parts; advection, longitudinal diffusion and reaction. For advection calculations the Holly-Preissman (1977) two-point four-order scheme was used whereas for longitudinal diffusion the Crank-Nickolson central difference method was used to discretize the equation. He applied this model to an irrigation scheme in Nepal and found the model capable of predicting sediment transport under equilibrium as well as non-equilibrium conditions.

The DORC (Design of Regime Canals) model was developed by Hydraulic Research Wallingford, for design of alluvial canals. The model provides a range of design methods together with procedures to predict alluvial friction and sediment transport. Alluvial canals can be designed by the regime, tractive force or rational methods. The regime methods used in the model are the Lacey and the Simons and Albertson method. Among the rational methods the Chang (1985) and Bettess et al., (1988) methods are included. Also the rational and regime methods can be combined in the model (Hydraulic Research (HR) Wallingford, 1992).

Belaud (1996) developed a sediment module of the hydrodynamic Simulation of Irrigation Canals (SIC) Model. The sediment module can simulate the sediment transport in non-equilibrium and unsteady state flow conditions. Belaud (2002) used this model in simulations of sediment transport in an irrigation system in Pakistan and proposed a number of measures for maintaining equity in silt effected irrigation systems.

3.5.   Sediment management in irrigation canals

Sediment control strategies start with the selection of a proper point for the diversion and the choice of appropriate structures at river intakes in order to prevent unwanted sediment entry into the irrigation canals. Then the entered sediments into the canals are ejected through different means, by structures or sometimes sediments are deposited in the oversized canal sections, settling basins, at the head of the canals and then periodically removed. Further, the canals are so designed that the hydraulic conditions during canal operation allow neither sediment deposition nor scouring in the canal prism. The offtaking structures are designed for maximum withdrawal of the sediments from the parent canal depending upon the command areas. Then the canal operation is planned in such a way that either of the phenomena is inhibited. Further details on these aspects are given in the subsequent sections.

3.5.1   Sediment control at intakes

Selection of point of diversion

Careful selection of the point of diversion is very important in reducing sediment entry to the irrigation canals. In general, the outside or concave side of the curve of a river has been proved the best location for river intake structure as shown in Figure 3.1. Because the heavy bed load is swept towards the inside of the curve due to spiral flow American Society of Civil Engineers (ASCE) (1972). This principal of bed load sweep was employed to the Sukkur Diversion Dam on the Indus River in Sindh Province, Pakistan, and bed sediment load was deflected away from the canal heading and diverted over the dam. The principal is also used by the United States Bureau of Reclamation (USBR) at number of its diversion sites (ASCE, 1972).

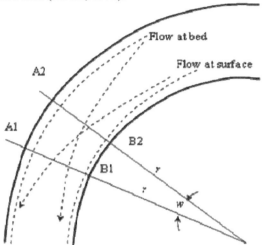

Figure 3.1. Schematic diagram of flow in curved channels (ASCE, 1972)

Angle of diversion

The angle of diversion between the direction of flow in the parent channel is generally called "Angle of Diversion" and sometimes "Angle of Twist". The higher velocity of surface water requires a greater force to turn it than the slower moving water near the bed. Consequently, the surface water tends, by its higher momentum, to continue with the parent stream; while the slower moving water near the bed, that carries the greater concentration of sediment, tends to flow into the diversion channel. Therefore, the diversion channel receives the sweep of the bed load, which flows from the outside to the inside of a curve, because for any angle of diversion the diversion takeoff is, in effect, on the inside of the curve created by the diversion. The most commonly used angle of diversion is 90°, which is definitely not the proper one (American Society of Civil Engineers, 1972). The use of an angle of diversion between 30° and 45° (from downstream side) is recommended as shown in Figure 3.2.

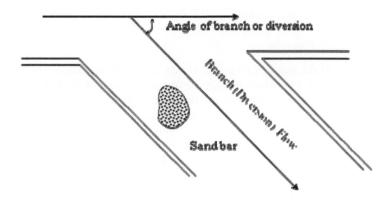

Figure 3.2. Schematic layout of desired angle of diversion (ASCE, 1972)

3.5.2   Sediment diverters (silt excluders)

A sediment diverter is a device or structure arrangement at a canal headwork which is designed to prevent the greater part of the stream sediment from entering the canal.

Training walls are the curved walls of the intake channel which create a curve in the flow artificially in which the helicoidal currents sweep the bed load to the inside of the curve and away from the headgates.

Guide banks used in some diversion works are curved banks which are designed to perform the same function as the curved training walls, that is, to divert sediment away from the canal intake. The guide banks have been extensively used on the diversion structures in India and Pakistan. Kotri Dam in India used guide banks and central islands to provide a concave curvature on each side of the river.

Pocket and divider walls are constructed upstream from a dam so as to form a pocket in front of the canal intake. The wall divides the stream flow as it approaches the dam, where the dam has a gated structure. The main function of the divider wall is to form a sluicing pocket upstream from the sluice gate to produce a ponding area of low velocity in which the sediment will deposit rather than enter the canal headwork. The sediment thus deposited in the pocket, is subsequently scoured out by turbulent flow through the pocket when the sluice gates are opened.

Sand screens that are used extensively in Egypt and India, are low barricade walls. They skim the clearer water off the top of the stream flow and direct the bed load away from the canal diversion.

Guide vanes produce localized helicoidal flow patterns that are similar to those generated naturally in flow around a curve. The bottom guide vanes direct the direction of flow in the lower part of the stream prism and surface guide vanes influence the direction of flow of the surface water as shown in Figure 3.3.

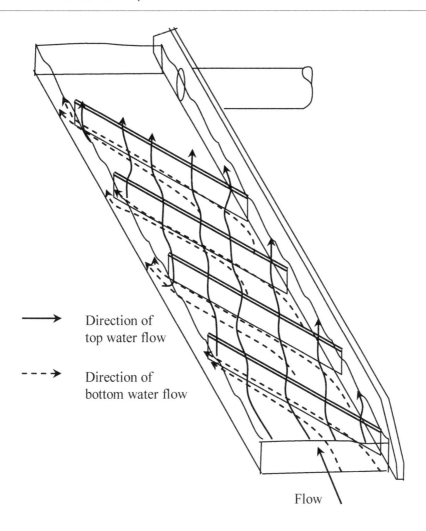

Direction of
top water flow

Direction of
bottom water flow

Flow

Figure 3.3. Bottom guide vanes (ASCE, 1972)

Tunnel type sediment diverters are composed of an upper and lower chamber through which the water to be diverted is routed. The clearer water flows through the upper chamber into the canal, while the sediment laden water flows into the lower channel and is passed back into the stream as shown in Figure 3.4.

3.5.3   *Sediment ejectors*

Sediment ejectors employ the same general principle of sediment removal as the diverters described previously except they are designed to remove sediment from the canal prism and are located downstream of the canal headgate.

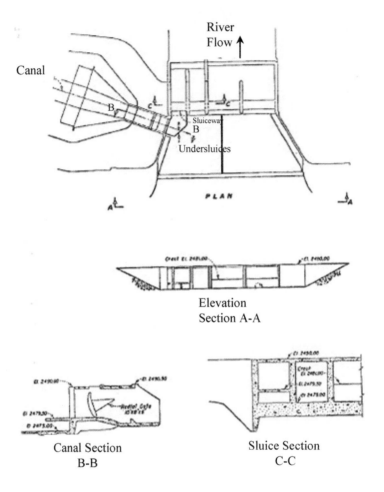

Figure 3.4. Undersluice tunnel type sediment diverter at Sargent Canal Headworks, Milburn Diversion Dam, Pick-Sloan Missouri Basin Programme, Nebraska, USBR (ASCE, 1972)

Tunnel type ejectors are most commonly used type of ejector in the large canals in India and Pakistan. This is very similar to the tunnel-type diverter, except that it is located in the canal prism to intercept and remove the residual canal bed load as shown in Figure 3.5.

Vortex tube ejectors consist of an open-top tube or channel. This open top tube is installed at an angle of 45° to the axis of flow in the canal headworks channel. The edge, or lip, of the tube is set level with the bottom grade of the canal. Water flowing over the opening induces a spiralling flow in the tube throughout its length. The spiral flow picks up the bed load and moves it along the tube to an outlet at the downstream end of the structure as shown in Figure 3.6.

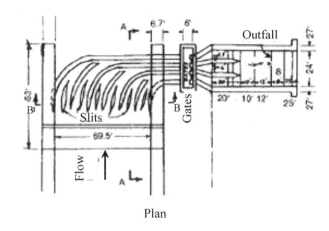

Plan

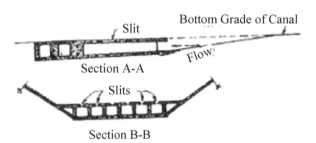

Section A-A

Section B-B

Figure 3.5. Tunnel Type sediment ejector at Salampur in Upper Bari Doab Canal, India (ASCE, 1972)

3.5.4   Settling basins

Settling basins are used to remove bed load and other sediment from the water at a canal headwork. These are generally placed in the canal just downstream from the headwork. The settling basin consists of an oversized section of the canal in which the velocity is low enough to permit the suspended particles to settle out, which are periodically removed.

3.5.5   Sediment control by canal design approach

Lined canals. Design of lined irrigation canals is relatively simple as there is not a certain restriction on higher or lower values of the flow velocities. As long as the

Manning's n or Chezy C is estimated correctly for the given lining material, the canal works as per design.

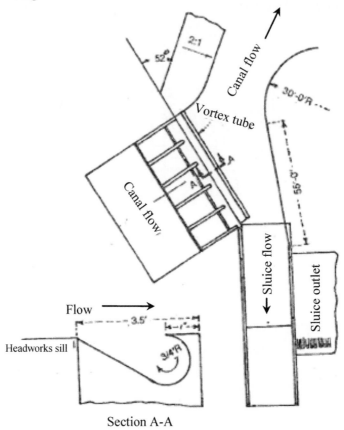

Section A-A

Figure 3.6. Vortex Tube sediment ejector – Hydraulic Model study, Superior Court Land Diversion Dam, Pick Sloan Missouri Basin Programme, Kansas, USBR (ASCE, 1972)

Unlined canals. The design of unlined canals needs a lot of care in order to make them stable. The purpose is to determine such values of depth, bed width, side slopes and longitudinal slope of the canal which produce a non-silting and non-scouring velocity for the given discharge and sediment load. A number of approaches have been developed to attain such conditions in the irrigation canals like empirical approaches developed in India and Pakistan (Indo-Pak) in which relationships between velocity, depth, hydraulic mean radius and slopes were determined in order to design stable cross-sections. Amongst those the Kennedy's silt theory and Lacey's regime theory are of particular importance (Ali, 1993).

Kennedy's regime concept

Kennedy was the first engineer to do research work on scientific lines. He was Executive Engineer, in charge of the Upper Bari Doab Canal (UBDC), published his work in 1895 through the Institution of Civil Engineers (Kennedy, 1895), London. His main conclusions were: (i) A regime channel is one which neither silts nor scours; (ii) the eddies generated from the bed support the silt in suspension, and therefore the silt-supporting power of the stream is proportional to the bed width and not the perimeter of the channel; (iii) introduced the "critical velocity", which means non-silting non scouring velocity. Then some the irrigation canals in the Indian Subcontinent were designed on the regime concept.

Lacey's regime equations

Lacey published his first paper through the Institution of Civil Engineers London in the 1929 (Lacey, 1929). Lacey studied the observations on channels recorded by researchers in various parts of the world and concluded that the non-silting, non-scouring velocity was a function of hydraulic mean depth, R, and not the depth, D, as shown by Kennedy. He developed equations for velocity, wetted perimeter, hydraulic mean radius and bed slope for stable canal design. Lacey's regime theory was widely adopted in the Indo-Pak subcontinent for stable canal design. In 1934 the Central Board of Irrigation, Government of India, adopted the Lacey equations as the basis for designing silt stable canals in alluvium (Ali, 1993). The hydraulic characteristics like hydraulic radius, wetted perimeter and bed slope of the canal were adjusted to achieve the non-silting and non-scouring flow velocity. The Lacey equations are given as (Ackers, 1992):

$$P = 4.84\sqrt{Q} \tag{3.1}$$

$$U = 0.625\sqrt{fR} \tag{3.2}$$

$$S_o = 0.0003 \frac{f^{5/3}}{Q^{1/6}} \tag{3.3}$$

$$f = \sqrt{2500d} \tag{3.4}$$

Where
P	= wetted perimeter (m)
R	= hydraulic radius (m)
d	= sediment size (m)
U	= mean velocity (m/s)
S_o	= bed slope (m/m)
F	= Lacey's silt factor for sediment size d
Q	= discharge (m³/s).

Maximum permissible velocity method

The method of maximum permissible velocity has been used in the United States of America for the design of unlined canals. The method ensures freedom from scour only i.e. implying that the canal is carrying clear water with no sediment load with cohesive or non-cohesive bed or banks, with little chance of silting. Maximum permissible velocities for various types of bed materials are given depending upon the bed material (Simons and Senturk, 1992).

Tractive force method

By definition the tractive force is the force acting due to flowing water on the particles composing the perimeter of the channel. It is a force exerted over a certain area and not on a single particle. This concept was first developed by Duboy (Ali, 1993). United States Bureau of Reclamation (USBR) also did some work on this concept (Raudkivi, 1990). This method is based on the consideration of the balance of forces acting on sediment grains and it is used for evaluating the erosion limits only. No sediment transport is considered. The tractive force method is suited if the water flow transports very little or no sediment (Breusers, 1993). The method assumes no bed material transport, it is only relevant for canals with coarse bed material and zero bed material sediment input (HR Wallingford, 1992). The permissible tractive force is the maximum unit tractive force that will not cause serious erosion of the material forming the channel bed on a level surface, and also can be termed as critical tractive force. The tractive force depends upon shear stress at the bottom and can be expressed as (Dahmen, 1994):

$$\tau = c\rho g y S_o \qquad\qquad (3.5)$$

Where

τ	= shear stress (N/m^2)
c	= correction factor depending upon the B/h ratio (B for canal width) and for wide canals c = 1
y	= water depth (m)
ρ	= density of water (kg/m^3)
S_o	= bottom slope
g	= acceleration due to gravity (m/s^2).

Hydraulic design criteria (HDC)

Ali (1990) stated that PRC/CHECKI consultants worked on the Irrigation System Rehabilitation Project, Phase-1 (ISRP-1) in 1985-1986 for the design of alluvial canals in IBIS. They used the Alluvial Channel Observation Program (ACOP) and Provincial Irrigation Departments to develop some relationships. They discarded the Lacey's design methods as arbitrary and out-dated. They proposed Hydraulic Design Criteria (HDC) and introduced silt traps and dredging as an integral part of canal design. They established some relationships for; (i) width computation, (ii) depth velocity relations, (iii) sediment transport capacity for the redesign of canal systems which have silted over a period of more than fifty year of their operation. The basic approach remained the same as was adopted by Lacey, but according to them, previously no attention was paid to the sediment input (through silty water entering the canal) and therefore problems occurred in situations with too low or too high sediment load.

3.5.6   Operation and maintenance of silt affected irrigation canals

Mahmood (1973b) studied sediment routing in irrigation canals in Pakistan with the goal to dispose of the incoming sediment load, particularly bed material load, with the water diversions. He found an increase in the total discharge consumption in conveyance losses along the system and therefore bed material concentration in the canal increased with distance. To overcome this discrepancy he suggested to increase the bed material load transport capacity of the system in the middle and lower reaches or allow sediment to accumulate in the middle and lower reaches of the system and periodically remove it by bed clearance. He found, with time, as the systems are operated, the bed material at each section tends to the size of load transported from upstream. He emphasized on withdrawal of bed material load with the irrigation diversions from a system as an important consideration in achieving a sediment discharge balance.

He emphasized that the disposal capacity of an irrigation system should be fixed according to the capacity of the farm watercourses rather than by the large conveyance canals in the system. Because in sediment routing through a canal network system, the problem of bed material transport becomes more critical as smaller canals are approached. This problem is most critical in tertiary canals.

Jinchi et al. (1993) presented a study on the sediment transport in an irrigation district which gets water from the Yellow River where sediment treatment is a key problem. They conducted a series of investigations on operation schemes for solving the problem in Bojili Irrigation System. They presented the effect of sediment deposition in the head reaches of the canals as a reduction in the water intake from the headworks because of raise in bed levels. They proposed that the large part of the sediments must be carried to some place far away from the headwork. They found that in irrigation canals sediment degradation and aggradation processes greatly depend on the hydrograph of water and sediment discharge. With some adjustments on the processes in a certain time interval, a best scheme, with which the sediment deposition in the upstream of the silting canal is the least, can be reached. Based on such an idea, it is possible to transport as much as possible sediments into a further area for deposition or for farmland usage, where sediment is taken as a natural resource and thus greatly improves the water diversion conditions in the irrigation systems with plenty sediment load input.

Bhutta et al. (1996) highlighted the flaws in the repair and maintenance policies of the government and of the Irrigation Department as well. They stated that the amount of money provided for maintenance works on the irrigation canals during the annual canal closure period was not sufficient. The method of assessment, by the Department of Irrigation, of the required maintenance works was not systematic and the limited availability of the budget was not utilized judiciously. Usually the desilting campaigns were done in the lower halves of the irrigation canals, which did not improve much the hydraulic performance of the canal. They found that if the desilting campaign was done in the upper two-thirds of the canal, it would greatly improve the water distribution in the canal.

Belaud and Baume (2002) reported that the planning of maintenance activities in surface irrigation systems was essential for optimal use of the annual credits. In many countries, the equity of the water distribution is largely affected by sediment deposition but the budgets do not allow the performance of all the necessary maintenance works and priorities must be defined by the irrigation agencies. A methodology based on numerical modelling was developed and illustrated on a secondary network in South Pakistan by the authors. Improvements on the current desilting procedure were proposed, but it was shown as well that the system could be designed differently in order to preserve the equity longer.

3.6.   Flow control in irrigation canals

In irrigation canals the flow control is necessary in order to properly regulate and distribute the flows to the offtaking canals. A variety of flow control methods is employed for this purpose. These flow control methods are the upstream control, proportional control, downstream control, Predictive (ELFLO) and Volume (BIVAL) control. The overall objective of the flow methods is to maintain a desired water level at specific locations in the canal reaches. The upstream control, proportional control and downstream control are of interest in this study and are described in the following section. Whereas in the remaining two methods, in Predictive Control, the water level is maintained at the downstream side of the canal reach (upstream of the structure) and in Volume control the water level is maintained in the middle of the canal reach.

3.6.1   Upstream control

In the upstream control system, the water level is maintained at the upstream side of the structure in order to feed the offtaking canals. This is the typical canal control system applied in the Indus Basin Irrigation System. Various types of overflow and underflow structures are employed in this system, like weirs, canal falls, flumes and the undershot gates, depending upon the requirements. Majority of the gated structures are manually operated. This system has fixed supply based approach. It is comparatively difficult in this system to operate it on demand based. For efficient operations, the calculation of response times is necessary in order to maintain the equity in water distribution.

3.6.2   Proportional control

In the proportional control system, the available water in the system is proportionally distributed to the offtaking canals. It has a very rigid approach and usually does not take into account the water demands in the command area, so it does not take into account the water shortages or surpluses in the command areas. Whatever flow is available in the system, is distributed proportionally to the users. This is the typical water distribution system in IBIS. It is, however, difficult to maintain equitable water distribution in this system due to canal's maintenance and sedimentation. Structures used in proportional control are more or less the same as in upstream control.

3.6.3   Downstream control

The upstream control and proportional control have fixed supply based operations and are not demand responsive. Further, significant time is consumed in canal filling after some period of closure, which affects the supplies to water users. In downstream control systems, these problems are solved. These systems are more responsive to water demands and have flexible operations. Water is immediately available to water users due to positive wedge storage in the canal reaches. The flow control structures are self operating and automatically respond to water withdrawal or refusal.

3.7.   Hydraulics of downstream control system

As clear from the name, flow control in the downstream control system is based upon the water levels downstream of the cross regulators. In Indus Basin Irrigation System the gated structures for water level regulation are called cross regulators. So depending upon water levels immediately downstream of the cross regulators the flow is released into the downstream reach of the canal. If water levels are lower at the downstream side of a cross regulator, more flow is released and if water levels are higher or close to the target water levels, less flow is released accordingly. Basically, in downstream control a target water level is maintained. Ankum (undated) describes the downstream control system as in downstream control the water level regulator responds to conditions in the downstream canal reach. Although it is a downstream water level control, it is called normally the downstream control (Ankum, undated).

Downstream control system is a demand based irrigation system. Ideally, in such systems the water is used according to the crop water requirements of the area. The offtakes are operated then according to the demand. It means outflow from the system varies quite frequently depending upon the crop water requirement of the command area. Therefore, a nominal storage is provided in the canal in order to provide immediate water supply to offtaking canals. In this way the time needed for canal filling can be avoided and the water is available whenever it is needed. Figure 3.8 presents the conceptual description of automatically downstream controlled irrigation canals.

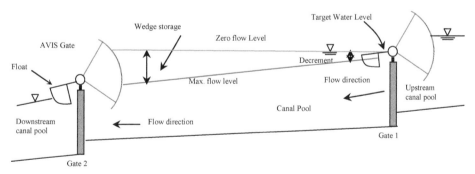

Figure 3.8. Schematic description of an automatically downstream controlled irrigation canal's reach

The Figure 3.8 shows two lines of water surface, the red line shows the water level at maximum discharge or design discharge, whereas the blue line shows the water level at zero flow. When offtaking canals are closed the water starts to store in the canal reaches. Then depending upon the available storage in the canal reach the water is stored. The inflow to the reach is regulated by hydro-mechanically self-operated AVIS and AVIO gates. Depending upon water levels in the canal and decrement setting of the gate the flow is then automatically regulated. Decrement is the difference between water levels at maximum flow and the water levels at zero flow. The water levels at zero flow are also called pond level (PL) or target water level. Figure 3.9 shows pond levels and 50% of pond level in Maira Branch Canal Pakistan.

A. Water level at high discharge   B. Water level at low discharge
Figure 3.9. Water levels close to the pond level during low flow

In Figure 3.9 there is a PL mark at the top of the gauge, which means pond level. At pond levels the flow is almost zero. Then below the pond level there is 50% PL, which means the water level at a discharge of 50% full supply discharge (FSD). Then below the 50% PL mark, there is the full supply water level mark, which is not visible in the picture because of higher water level. It shows that at low flows the water levels are higher and at high flows the water levels are lower. This raise or drop in water level operates the AVIS/AVIO cross regulators.

Figure 3.10 shows an AVIS cross regulator. It consists of many parts, but the float, counter weight and gate radius need description in this study. The float of an AVIS or AVIO gate is the main driving agent. As mentioned previously any reduction in the outflow from the canal causes a raise in water level in the canal. A raise in water level exerts a force on the float and lifts it up depending upon the decrement value. As the float is lifted up it brings down the gate leaf, which causes a reduction in the flow to the downstream canal reach. The counter weight balances the weight of the gate leaf, so it is easily operated by any change in water level.

The gate axis is set at the water level in the downstream side, i.e. the water level at zero-flow. Since the gate leaf is radial, all thrust on the gate leaf due to difference of upstream and downstream water levels, passes through the axis. The radial float is attached to the gate and receives the thrust of the water pressure. The weight of the gate is not relevant for the operation as its effect is neutralized by balancing. Thus the only torque that comes into play is due to the buoyancy of the float. The counterweight and the float are both ballasted during the installation process of the gate.

The downstream controlled irrigation canals have a horizontal berm in order to accommodate the positive storage wedge. The canal berm between two consecutive cross regulators is horizontal, means flat, and then in the downstream of a cross regulators the new bank level is defined according to the given headloss and hydraulic conditions in the canal, which again remains flat up to the next cross regulator and it goes on till the end of the canal. Figure 3.11 presents water levels in the canal at two different conditions, at high flow Figure 3.11.A and at low flow Figure 3.11.B. It can be seen that the canal berms are quite flat and easily accommodated the raise in water level due low flows.

Figure 3.10. An AVIS cross regulator in Maira Branch Canal, Pakistan

A. Water levels at high discharge   B. Water levels at low discharge
Figure 3.11. Water levels at high and low discharges

There are some merits and demerits of this irrigation system. This system is demand based and has quite flexible canal operations. The flow regulating structures are automatically operated. The water is immediately available to the offtaking canals, hence there are minimum operational losses. Due to self regulation of flow regulating structures the new steady state conditions are achieved quite efficiently. On the other hand this system is quite expensive because more earthwork is required to have the horizontal berms. This system becomes even more expensive if it is built in hilly terrain or steep sloping areas. As it is a demand based system, so its operation becomes quite complicated in case of water deficiency. Auto-functioning of the AVIS/AVIO gates needs to be checked periodically. Sometimes gate's setting itself or other factors affecting the hydraulics of the canal cause undue closing of the gates, which results in overtopping of the AVIS gates as shown in Figure 3.12. This overtopping can be very hazardous for the stability of the downstream canal reach and for the gate safety as well.

Figure 3.12. Overtopping of the AVIS gates in Maira Branch Canal, Pakistan

3.8. Design criteria

Downstream control systems are designed for maximum and minimum discharges at the same time because when the discharge is zero, the water level in the canal is higher than the water level at full supply discharge. So the canal needs to be designed to pass the maximum or design flow and for the minimum or zero flow and to store the additional water during low flows.

3.8.1. Designing for maximum flow

The automatically downstream controlled irrigation canals are generally designed for maximum flow on the uniform flow concept. The flow is said to be steady and uniform if flow characteristics such as depth, velocity, discharge, roughness and canal slope do not change in different cross sections of the canal length. The streamlines are considered rectilinear, parallel and the pressure distribution is hydrostatic under these conditions. In uniform flow the slope of the flow, of the channel bed and of the energy grade line remain same (Graf, 2003).

The resistance to flow in uniform flow plays vital role. Chaudhary (1994) describes as the component of weight of water in the downstream direction causes acceleration of flow, whereas the shear stress at the channel bottom and sides offer resistance to the flow. If the channel is long and prismatic then the flow accelerates for some distance until the accelerating and resistive forces are equal. From that point on, the flow velocity and flow depth remain constant. Such a flow, in which depth does not change with distance, is called uniform flow, and the corresponding depth is called the normal depth.

The two most widely used flow resistance formulae in open channels are the Chezy's formula, Equation 3.6 and the Manning's formula, Equation 3.7, for uniform flow. These formulae are given as:

$$U = C\sqrt{RS_f} \qquad\qquad\qquad (3.6)$$

$$U = \frac{1}{n}R^{2/3}S_f^{1/2} \qquad\qquad\qquad (3.7)$$

Where
U	= mean flow velocity (m/s)
C	= Chezy coefficient for channel resistance (m$^{1/2}$/s)
R	= hydraulic radius (m)
S_f	= bed slope
n	= Manning's roughness coefficient (s/m$^{1/3}$)

3.8.2. Designing for minimum flow

The downstream controlled irrigation canals are also designed for minimum flows at the same time, which can be zero as well. When the flow in the canal is forced to reduce from full supply discharge or from the maximum flow, the water starts to store in the canal and water levels rises up. To accommodate this storage, the downstream controlled irrigation canals are provided with level embankments for wedge storage.

Wedge storage

As mentioned in the previous paragraph that when there is less discharge in the canal than the design discharge, then the additional water is stored in the canal prism. This storage is provided for immediate water supplies to offtaking canals when they are opened again or need more water again.

The storage in the canal is $\forall_{min}$ in order to meet the sudden increase in demand and also to ensure stability of the system by avoiding oscillations. The minimum volume, $\forall_{min}$ of the storage wedge can be determined from the following equation (Ankum, undated):

$$\forall_{min} = \tfrac{1}{2}\,Q(T_u + T_w) \qquad (3.8)$$

Where
  $\forall_{min}$ = minimum storage volume (m^3)
  T_u = travel time upwards (seconds)
  T_w = travel time downwards (seconds)

T_u and T_w are determined as:

$$T_u = L/(\sqrt{gd} - U) \qquad (3.9)$$

and

$$T_w = L/(\sqrt{gd} + U) \qquad (3.10)$$

Where
  L = length of the reach (m)
  g = acceleration due to gravity (m/s^2)
  d = hydraulic depth (m)
  U = flow velocity (m/s)

The available volume of positive storage wedge, $\forall_{dyn}$, is determined as;

$$\forall_{dyn} = \tfrac{1}{2}\,W\,\Delta y\,L + \tfrac{1}{3}\,m(\Delta y)^2\,L \qquad (3.11)$$

The effect of decrement ε on storage wedge, $\forall_{dyn}$ is given as:

$$\forall_{dyn} = \tfrac{1}{2}W\Delta yL + \tfrac{1}{3}m\Delta yL + W\varepsilon L + m(\varepsilon^2 + \varepsilon\Delta y)L \qquad (3.12)$$

Distance between cross-regulators

The minimum distance L_{min} between two regulators, depending upon minimum volume required for the storage wedge, can be determined from the following equation. A minimum distance L_{min} for minimum volume V_{min} is usually in the range of 1 km to 3 km when decrement is not included in the calculations. This L_{min} even becomes smaller when certain decrement is assumed. Generally, the distance between two cross regulators is determined on the basis of cost. If a small distance is chosen then more gates are required, which are usually expensive. In case of large distance, the dynamic volume, V_{dyn} becomes much larger than the V_{min} which needs more earthwork for hydrodynamic stability of the canal reach.

$$L_{min} = \frac{-\left(\dfrac{W}{2}+\varepsilon\right)+\sqrt{\left(\dfrac{W}{2}+\varepsilon\right)^2+\dfrac{4}{3}m\left(\dfrac{\sqrt{g}v^{\frac{1}{2}}W^{\frac{1}{2}}Q^{\frac{1}{2}}}{gQ-v^3W}-W\varepsilon-m\varepsilon^2\right)}}{\frac{2}{3}ms} \qquad (3.13)$$

Where

n	$=$ Manning's roughness coefficient (s/m$^{1/3}$)
Q	$=$ the maximum discharge (m^3/s)
V	$=$ the maximum velocity of water (m/s)
W	$=$ the width of the water level during the maximum discharge (m)
m	$=$ the side slope
s	$=$ water surface gradient during maximum discharge
ε	$=$ decrement (m)
L	$=$ length of the canal reach between the regulators (m)
y	$=$ the water depth during the maximum discharge (m)
B	$=$ bed width (m)
V_{dyn}	$=$ the in-canal storage, i.e. the volume of the canal reach between the old and the new water level (m^3)
Δy	$=$ maximum change in water level (m)

Location of the offtakes

The better location for an offtakes is just downstream of a water level regulator, where the variation Δy in water level is limited to the decrement ε of the gate. Offtakes located between the head end and the tail end regulator experience larger water level variations. The water level variation, Δy, can be estimate as, $\Delta y = \varepsilon + xs$, with the decrement ε of the head-end gate, the distance, x, to the head-end gate and the (energy) gradient s.

3.8.3. AVIS and AVIO Gates

Downstream control canals are equipped with AVIO and AVIS gates and are used to control water levels at the canal headworks and in canal reaches. The gates maintain a constant water level at their downstream side (upstream end of a canal pool). These are hydro-mechanical self-operating gates. The AVIS and AVIO gates are similar gates. The name 'AVIS' has a French background: AV is from "*aval*", means downstream, and S is from "*surface*", whereas in "AVIO" the letter "*O*" is from "*Orifice*". It illustrates that the AVIS gate operates at a free surface flow and AVIO gates operates under orifice conditions as shown in Figure 3.13. Further, two types of AVIS/AVIO gates are available: the "High Head" type and the "Low Head" type. The High Head gates have a more narrow gate than the Low Head gates with the same float. The choice between the open type (AVIS) and the orifice-type (AVIO) is solely determined by the maximum headloss likely to occur between the upstream and downstream-controlled levels.

A. A pair of AVIS gates   B. A pair of AVIO gates
Figure 3.13. AVIS and AVIO gates in action

3.9.   Flow control algorithms

Hydraulic efficiency of either upstream or downstream control irrigation canals can be improved with canal automation. The proper development or selection of canal control algorithms plays pivotal role in achieving this objective by proper adjustment of flow regulating structures. A control system is an elementary system (algorithm software and hardware) in charge of operating canal cross structures, based on information from the canal system, which may include measured variables, operating conditions and objectives. Boundaries of the control system are output of the sensors placed on the canal system, and input to the actuators controlling the cross structures. Malaterre (1998) classified the canal control algorithms on the basis of considered variables, logics of control and design technique use for a control system.

The considered variables are the controlled, measured and control action variables. *Controlled variables* are target variables controlled by the control algorithm for example water levels at the upstream or downstream end of a pool, discharge at a structure and volume of water in a pool. *Measured variables,* also called inputs of the control algorithm, are the variables measured on the canal system. Examples are water level at the upstream end of a pool, downstream end of a pool, at an intermediate point of a pool, flow rate at a structure, and setting of a structure. *Control action variables*, also called outputs of the control algorithm, are issued from the control algorithm and supplied to the cross structures' actuators in order to move the controlled variables toward their established target values. They are either gate positions or flow rates.

For purposes of describing and classifying canal control algorithms, an *Input/Output (I/O)* structure is used, consisting of the number of inputs and outputs considered by control algorithm. The *I/O* structure of a control algorithm is described as *"nImO"* when it has n inputs (measured variables) and m outputs (control action variables). Special names are give to the *I/O* structure in specific cases for example *Single Input Single Output (SISO)*, if $n = m = 1$; *Multiple Inputs Single Output (MISO)*; if $n > 1$ and $m = 1$; and *Multiple Inputs Multiple Outputs (MIMO)*, n and $m > 1$. This structure has influence on the techniques that could be used for the design of the algorithm.

The *logic of control* refers to the type and direction of the links between controlled variables and control action variables. The control algorithm uses either feedback control (FB, also called closed-loop control), feedforward (FF, also called open-loop control) or a combination (FB + FF).

In *Feedback control algorithm*, the controlled variables are measured or directly obtained from measurements. Any deviation from the target is fed back into the control algorithm in order to produce a corrective action that moves the controlled variables towards the target values, Figure 3.14.A. Perturbations, even if unknown, are taken into account indirectly, through their effects on the output of the system. In control theory this concept is essential because it links a control action variable to a controlled variable. Feedback controls can be applied to all the controlled variables, discharge, water level and volume.

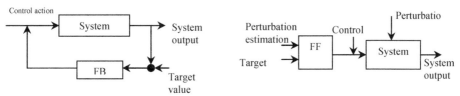

A. FB control system   B. FF control system
Figure 3.14. Feedback and feedforward control systems

Examples of feedback control in discharge are GPC (Sawadogo et al., 1992b; Rodellar et al., 1993), in water level are *AMIL, AVIS, AVIO* gates, *Little-Man, EL-FLO* (Shand 1971) and *PID*.

In a *feedforward control algorithm*, the control action variables are computed from targeted variables, perturbation estimation, and process simulation as shown in Figure 3.14.B. Feedforward control usually improves control performance when few unknown perturbations occur in the canal system. The feedforward control can compensate inherent system time delays by anticipating users' needs. Feedforward controls can be applied to all the controlled variables; discharge, water level and volume.

3.9.1. Supervisor Control and Data Acquisition System

Supervisory Control And Data Acquisition (SCADA) is a real time discharge monitoring and control system, which collects the data on actual water levels and automatically controls the discharge supplies. A setpoint is established in the canal and any deficiency/excess to this setpoint is automatically adjusted, based on the choice of the automatic discharge controller. A sensor is installed at a required location in the canal reach and is connected electronically to the Master Control Panel (MCP), from where the control actions are taken according to the difference between target and actual water levels. The MCP is supported by a Hydraulic Power Pack to operate regulation valves. The discharge controllers in the SCADA system can be of any type as mentioned above. In a canal control, usually PI (Proportional Integral) controllers are provided. A Human Machine Interface (HMI) is provided to exchange information with the operator and to enter new desired values.

3.9.2. Proportional Integral Derivative (PID) control

A Proportional Integral Derivative (PID) is a feedback controller. It compares the system output with a target water value and instructs the system about new output to meet the target values. In case of canals, the target values can be water levels or discharges. The *PID* is a combination of *P, PD* or *PI* controllers.

The *P* controller is the simplest continuous controller. Any deviation *e* at a moment *t* actuates the regulator and is proportional to the difference e between measured water level and the setpoint (the target water level). The intensity of the controller action is given by the proportional gain, K_p. A low absolute value of K_p leads to damped gate reaction and a high value leads to a strong reaction of the regulator and may cause instability. The action from the *P* controller can be supported by applying a damping effect to the output by a *D* controller. The *D*-controller (Derivative controller) is added to the *P*-controller, to create a PD-controller. Its function is to anticipate the future behaviour of the controlled variable (water level or discharges) by considering the rate of change. The *D*-controller avoids any rapid increase or decrease in the water levels or discharges. An *I*-controller always forces the system output to the setpoint. The integral gain factor, K_i, inserts a memory of deviations e of the past, by taking the sum of all deviations up to the present time. Finally the *PID* is the combination of these three controllers. It became available commercially in the 1930s. The first computer control applications in the process industries entered into the market in the early 1960s (Seborg et al., 1989).

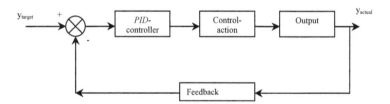

Figure 3.15. Block diagram of PI controller

Mathematically a *PID* controller can be written as Equation 3.14:

$$Controller\ output = K_p \left(e(t) + \frac{1}{T_i} \int_0^t e(\tau)d\tau + T_d \frac{de}{dt} \right) \tag{3.14}$$

Where

K_p	= proportional gain
T_i	= integral action time (seconds)
T_d	= derivative control action
e	= system error

4. Sedimentation in Tarbela Reservoir and its effects on Pehure High Level Canal

4.1. Sedimentation of reservoirs in the world

"The ultimate destiny of all reservoirs is to be filled with sediments" (Linsley et al., 1992), the question is how long it takes to be filled up. Sedimentation in reservoirs is a natural phenomenon and cannot be avoided. When a river enters into a reservoir, the flow velocities drop quite low and the transport capacity of the river falls considerably. Owing to this almost all of the coarse material is shed by the river in the beginning of the reservoir. Only fine material remains suspended in the water and moves forward. Some examples of the reservoir are presented in Table 4.1. Batuca and Jordaan (2000) mentioned that some reservoirs just filled up with sediments within a period of two years of the impoundment, whereas they also presented some reservoirs, which took 70 to 80 years in filling up with sediments after the impoundment. Sedimentation in a reservoir depends upon the size of the reservoir, incoming sediment load and the reservoir operation strategy.

4.2. Tarbela Dam

Indus River is one of the world's highest sediment carrying rivers. The main factors causing high sediment load in the river are mainly the moving glaciers crushing rocks, comparatively young nature of geologic formations in the river basin and steep slopes in the area. When such rivers are impounded, severe sedimentation takes place in the reservoir. Sedimentation in the reservoir then causes some serious problems in the reservoir operation, amongst which the major ones are the loss of water storage capacity, risk of clogging of low level tunnel outlets, abrasive action of sediment laden flows on power turbines causing increase in maintenance costs.

Tarbela Dam (Figure 4.1) is located in Haripur District of Khyber Pakhtunkhwa Province. The dam was constructed in 1974 under replacement works undertaken after the Indus Waters Treaty, 1960. It is one of the world's largest earthfill dams. It was planned primarily as a storage reservoir for irrigation but it also contributes substantially to hydropower generation and flood control. It plays a crucial role in the economy of Pakistan by providing 40% of country's water storage for irrigation and 35% of the country's energy requirements (Hydraulic Research (HR) Wallingford, 1998). Salient features of the dam are given in Table 4.2.

The dam has been built on the Indus River, around 100 km away from Islamabad. The Indus River originates from 5,180 m above mean sea level (MSL) from Lake Mansrowar, located in the Kailash ranges in Himalaya. The basin area of Indus at Tarbela is 169,600 km², encompassing seven world's highest peaks and seven largest glaciers. Inflow to the reservoir consists 90% of snow and glaciers melt. (Haq and Abbas, 2006).

Figure 4.1. Tarbela Dam on Indus River

Table 4.1. Sedimentation of reservoirs in the world

Country	Number of reservoirs	Silted volume (MCM)		Silting percentage (%)		Silting duration (years)	
		from	to	from	to	from	to
Algeria	11	0.6	52	3	83	2	71
Austria	13	0.1	19	0.2	72	8	25
China	7	3	3,400	10	99	1.5	16
Ethiopia	3	2.3	50	3	75	3	10
France	3	7	28	6	50	12	40
Germany	3	1.2	2.4	2.9	100	7	18
India	11	5.2	495	0.1	62	7	37
Japan	4	4	76	23	96	8	22
Spain	5	17	50	6	97	9	38
Former USSR	13	1.1	313	0.4	99	4	24
USA	54	0.1	5138	0.5	60	4	70

(Source: Batuca and Jordaan, 2000)

About 94% of the Indus Basin at Tarbela lies outside the monsoon region with scanty rainfall and precipitation mainly in the form of snow. Only, a small portion of 10,400 km^2 falls in the active monsoon zone, which lies immediately above Tarbela Dam. The annual average precipitation in this area ranges from 800 to 1,500 mm. The moving glaciers crush rocks on their way and leave behind a lot of sediments when they melt, which is carried by the river due to the steep gradient.

Table 4.2. Salient features of Tarbela Reservoir

Item	Parameter	Description
Reservoir	Length (km)	96
	Maximum depth (m)	137
	Area (hectare)	24,300
	Gross storage capacity (BCM)	14.3
	Live storage capacity (BCM)	11.9
	Dead storage (BCM)	2.4
Embankments		
Main embankment dam	Length (m)	2,743
	Crest Elevation (m+MSL)	477
Auxiliary dam 1	Length (m)	713
	Crest Elevation (m+MSL)	472.4
Auxiliary dam 2	Length (m)	292
	Crest Elevation (m+MSL)	472.4
Service spillway	No. of gates (15.2 x 18.6 m)	7
	Discharge capacity (m^3/s)	17,400
	Ogee level (m+MSL)	454.7
Auxiliary spillway	No. of gates (15.2 x 18.6 m)	9
	Discharge capacity (m^3/s)	22,500
	Ogee Level (m+MSL)	454.7
Tunnels		
Right Bank Tunnels	Length (m)	731 to 883
	Tunnel Diameter u/s of gate shafts (m)	13.7
	i). Tunnels 1, 2, 3 (m)	13.3
	ii). Tunnel 4 (m)	11
	Intake elevation of tunnel 1 & 2 (m+MSL)	373.4
	Intake level of tunnel 3 & 4 (m+MSL)	353.5
	Intake level of Gandaf Tunnel (m+MSL)	393.2
Left bank tunnel	Length	1,120
	Tunnel Diameter u/s of gate shaft (m)	13.7
	Intake level of tunnel 5 (m+MSL)	362.7
Power plans		
Power plant	Units 1-10 @ 175 MW each	1,750
	Units 11-14 @ 432 MW each	1,782

4.3.   Sedimentation in Tarbela Reservoir

4.3.1   Water availability at Tarbela Dam

Figure 4.2 presents the total annual water availability at Tarbela Dam from 1962 to 2006. The average annual flow at Besham Qila is 76 BCM, with a standard deviation of 10 BCM. The maximum and minimum annual inflows to the Tarbela Reservoir have been recorded as 91 BCM and 59 BCM in the years 1994 and 2001 respectively after dam construction. Whereas the maximum annual flow recorded at Tarbela has been 102 BCM in 1973.

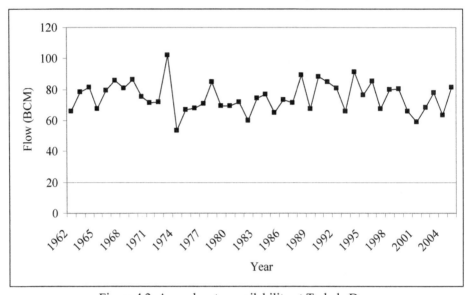

Figure 4.2. Annual water availability at Tarbela Dam

Water availability is distributed in such a way that about 90% of water in Indus River is available in summer from May to September as shown in Figure 4.3, mainly from snow and glacier melt and rest of the water is available in winter months October to April. The average annual flow rate becomes 2,400 m^3/s in Indus River at Besham Qila.

4.3.2   Sediment inflow to Tarbela Reservoir

The annual sediment load entering into the Tarbela Reservoir is given in Figure 4.4. The sediment load is measured at the Besham Qila rim station, which is about 100 km upstream of the dam, and is increased by 10% in order to accommodate the sediment load from the ungauged river basin between dam and the Bisham Qila rim station. The average annual sediment load coming to Tarbela Reservoir has been estimated as 203 MT. The maximum and minimum loads of sediment have been observed as 426 and 73 MT in the years 1990 and 1993 respectively.

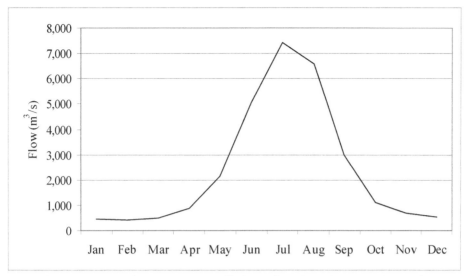

Figure 4.3. Monthly average water discharge in Indus River at Besham Qila

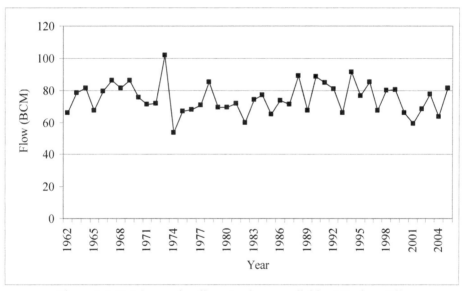

Figure 4.4. Total annual sediment volume available at Besham Qila

Like the water flow, more than 90% of the annual sediments come in the summer months, May to September, as shown in Figure 4.5. It is mainly due to glacier melting in summer and erosion of rocks and steep slopes in the basin area.

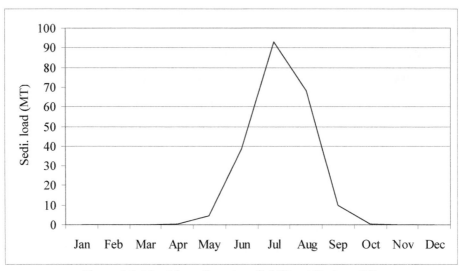

Figure 4.5. Monthly sediment availability at Besham Qila

4.3.3   Tarbela Reservoir operation

Tarbela Dam's main objective is to supply water for irrigation purposes to the Indus Basin Irrigation System (IBIS). Dam operation is decided primarily on the basis of basin's irrigation water requirements and hydraulics of the reservoir in order to control the advance movement of the sediment delta. An operational rule curve is prepared, usually, every year for dam operation in order to fulfil water demands, and somehow controlling sediment transport in the reservoir. In order to control delta's forward movement the minimum pool level is increased, usually, every year. At the time of completion of the dam, the minimum pool level was 396.2 m+MSL, which now has been raised to 417.2 m+MSL.

4.3.4   Sediment trapping in the reservoir and status of storage capacity

Large amounts of sediment settle every year into the reservoir. The trap efficiency of the reservoir is about 95% and on average every year about 192 MT sediments deposit in the reservoir. Figure 4.6 gives an overview of accumulated sediment deposition in the reservoir since the damming of the river flows. This accumulation of sediment reduces the storage capacity of the reservoir. Haq and Abbas (2006) reported that the storage capacity of Tarbela Reservoir was being lost at a rate of 0.132 BCM annually in previous years.

The latest data on sediment deposition was collected from the Project Monitoring Organization (PMO), WAPDA and assessed the effects of sediment deposition on the reservoir's storage capacity. Table 4.3 presents the latest status of the gross and live storage capacity of the reservoir. It was observed that by September 2008 the reservoir had lost its gross storage capacity by about 31.3%, whereas loss in live storage and dead storage was 29.2% and 66.2% respectively.

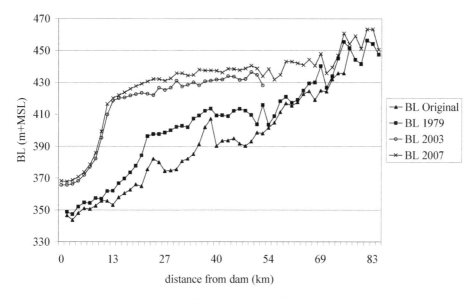

Figure 4.6. Pattern of sediment deposition in the Tarbela Reservoir

Table 4.3. Status of storage capacity in September 2008

S. No.	Status of storage	Quantity
1	Minimum reservoir level (m+MSL)	417.3
2	Original gross storage capacity (BCM)	14.3
3	Original live storage capacity (BCM)	11.9
4	Dead storage at elevation 396.2 m+MSL (BCM)	2.4
5	Gross storage by September 2008 (BCM)	9.8
6	Live storage by September 2008 (BCM)	8.5
7	Dead storage by September 2008 (BCM)	0.8
8	Loss in gross storage up to 2008 at elevation 417.2 (m+MSL) (BCM)	4.5
9	Loss in live storage up to September 2008 at elevation 417.2 (m+MSL) (BCM)	3.5
10	Loss in dead storage (BCM)	1.5
11	Capacity below min. reservoir level 417.3 (m+MSL) (BCM)	1.7
12	Sediment volume by 2008 (BCM)	4.5
13	Loss in gross storage (%)	31.3
14	Live storage loss (396.22-417.25 m+MSL) (%)	29.2
15	Loss in dead storage (%)	66.2

(Source: Project Monitoring Organisation, WAPDA)

4.3.5   Accumulated sediment deposition in Tarbela Reservoir

Due to massive settling of sediments the bed levels of the reservoir have raised even up to an extent of 70 metres since the start of the impoundment. Figure 4.7 presents cross-sections of the reservoir at four different sites along the reservoir, at 1 kilometre (km),

28.4 km, 37.6 km and 39 km. The cross-section at 28.4 km shows enormous deposition, where the maximum accumulated depth of sediments has been measured as 60 metres. The average and maximum depths of sediment accumulation have been given in Figure 4.8 as bed levels of the reservoir in m+MSL.

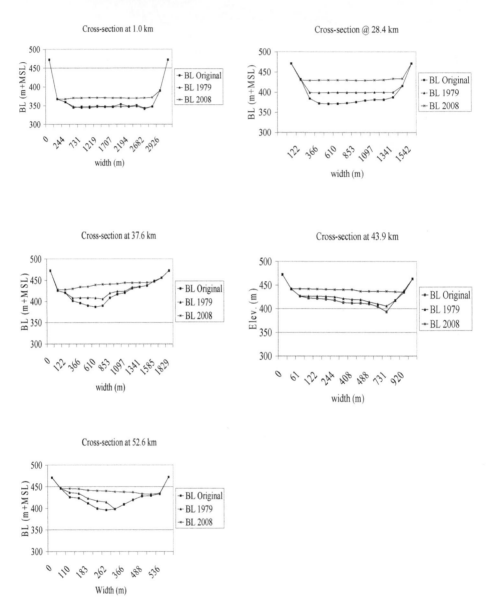

Figure 4.7. Cross-sectional view of sedimentation in the Tarbela Reservoir

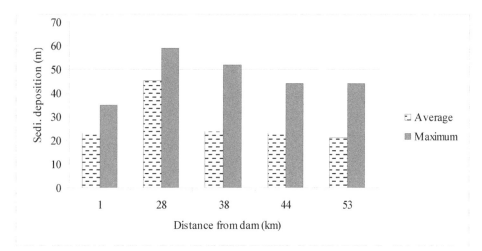

Figure 4.8. Average and maximum deposition at various locations in Tarbela Reservoir

4.4.   Development of sediment delta in a reservoir

4.4.1   Siltation process in reservoirs

The silting process in a reservoir generally starts in the upper shallow waters due to backwater effects. In this region, the water sediment flow is drastically decelerated and consequently the bed load transport is slowed down or even diminishes and the suspension carrying capacity of the flow is reduced as well. There are three main types of sediment deposition in the reservoirs, Backwater deposits, deltaic deposits, and bottomset deposits (Batuca and Jordaan, 2000). Backwater deposits are developed in the upper backwater reaches of reservoirs, and make a transition from natural rivers to delta formation in the basin. These deposits normally consist of coarse sediments and may extend up to and even above the reservoir's normal water level. Bottom-set deposits are developed in the main body of the reservoir basin and close to the dam wall and consist of fine and very fine sediments (silt and clay) transported by turbulent suspension and by turbidity currents.

4.4.2   Formation of a delta

Backwater deposits in the upper reaches of the reservoir are transformed into delta deposits when they move forward towards the dam. Delta deposits are developed within the reservoir basin and may extend both along and across the basin and normally consist of a non-homogeneous mixture of sediments. Delta deposits consist of the coarsest material of the sediment load, which is rapidly deposited at the inflow zones. It may consist of entirely coarse sediments (d > 0.062 mm) and may also contain a large fraction of finer material such as silt (Morris and Fan, 1997). As sediment deposition within the reservoir basin continues, the delta deposits progress, moving downstream towards the dam wall and growing in size and volume.

Depending upon the reservoir hydrology, hydraulics, sediment transport, size and configuration, the speed of progress of the delta is variable from 50 to 200 m per annum, as recorded at the Watts Bar Reservoir in the United States and at the Anchicaya Reservoir in Colombia, but an extreme value of 5,300 m per annum was observed in Lake Mead in the United States. The total travel distance covered by these deltaic deposits varies considerably as well, from 2 km (Anchicaya Reservoir in Columbia and Kulkuan Reservoir in China), to 17-18 km (Guernsey Reservoir in the USA and Izvoru Muntelui Reservoir in Romania), to 30-35 km (Welbedacht Reservoir in South Africa), and even to 60-80 km (Lake Mead in the USA and Bhakra Reservoir in India) (Batuca and Jordaan, 2000).

There are two distinct zones in the deltaic deposits, the top-set and the fore-set. The top-set and fore-set are separated by a pivot point as shown in Figure 4.9. Top-set deposits are formed by settling of bed-load, which is transported and deposited in the headwaters of the reservoirs. The top-set deposits are increased in size, depth and volume when the delta progresses towards the dam wall due to sediment accumulation. Fore-set deposits are formed at the head of the delta formation, by accumulation of all sediment carried in by the river, except for the fine sediments which settle in the bottom-set deposits.

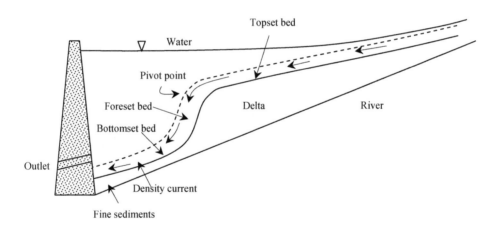

Figure 4.9. Typical sediment delta in a reservoir

4.4.3   Factors affecting sediment delta movement

Reservoir deltas reflect the interaction between inflowing streams and deposited bed material. The fluvial process affecting deltas is similar to that of an alluvial stream. The bed continuously adjusts in the changing water and sediment inflows and changing reservoir water levels. Factors affecting the formation of a sediment delta are mainly the water-sediment inflow regime, morphological characteristics of the reservoir and its general hydraulics and are influenced by dam structure and reservoir operation.

The reservoir operating rules have a large influence on sediment deposition accumulation pattern. Any change in operation rules can be used to alter the sediment deposition pattern. In a pool subject to wide fluctuations in water levels, delta deposits at high pool levels can be scoured and redeposited near the dam. Reservoir's pool level has great effect on the delta front advancement. If the reservoir is operated near top of the conservation pool at most of the time, the elevation of top of the conservation pool becomes the elevation of the pivot point. If there are frequent fluctuations and there is a deep established inflow channel, a mean operating pool elevation is used to establish the pivot point. If the reservoir is emptied every year during flood peak flows for sediment sluicing, there will be no pivot point i.e. the top-set and front-set deposits will merge to an intermediate slope (Batuca and Jordaan, 2000).

The delta form is affected by the lateral confinement of the inflow. If the valley is narrow the form is straight elongated. If the stream flows into the body of water that is substantially wider than the stream the delta grows both in length and width. The energy grade line over the delta is higher than that of adjacent water in the reservoir. The initial shape of the delta is elongated in the flow direction. The energy grade line is higher than that of adjacent water in the reservoir. As the flow slows down over the delta, and looses its momentum, the lateral energy gradient becomes significant and causes the flow to escape sidewards. The stream may swing to one side or bifurcate. The stream then carries sediment sidewards and adds to the delta laterally (Raudkivi, 1993).

4.4.4   Reworking of the delta

Large volumes of sediment deposits in delta areas during periods of flood inflows can be reworked during periods of subsequent drawdowns. Mainly coarse sediments are deposited in the head of the reservoir by backwater effect during high discharges, forming a delta. In Tarbela Reservoir, in the beginning of the flood season, the reservoir level is maintained at minimum pool level from May to June. The first high flows ranging from 1,500 m^3/s to 5,000 m^3/s are used to rework and flush part of sediment which is deposited in the year before. The deposits are laid down in the upper reaches of the reservoir when the reservoir level rapidly rises in June, when filling starts (Sloff, 1997). During minimum pool level the incoming floods erode a flushing channel in the deltaic deposits. The flushing channel gradually increases in width by bank-erosion processes during that period. For instance channel widths increasing from some 400 to 1,400 m were reported in 1981 by discharges up to 5,000 m^3/s. Highly erosive flows with suspended-sediment concentrations of about 20 times the inflow concentration, are moving the upstream deposits to the delta front. For instance in 1981 the top of the foreset slope advanced 4.8 kilometres during the flushing period.

4.5.   Analysis of the sediment delta in Tarbela Reservoir

As mentioned above, the delta in the Tarbela Reservoir is not stable and advancing towards the dam, as shown in Figure 4.6, mainly due to dam operation and changes in pool levels. Various studies found that the delta front is moving with a speed ranging from 0.6 to 1.0 km per year. For instance, Sloff (1997) found that depending on the reservoir operation the average rate of advance of the delta to the dam is about 600 m in a year (Sloff, 1997). Whereas Haq and Abbas (2006) found that the delta is located at a distance of 10.6 km upstream of the dam and is advancing at a rate of 1 km per year. In Figure 4.6 the changes in pivot point of the delta have been compared based on long term data. The pivot point location in 1979 was compared with the location in 2003 and found that the delta had moved a distance of 11.6 km. Whereas the comparison between pivot point locations in 2003 and 2007 show a difference of 2.2 km. In both cases the delta front's advancing rate approximately comes to 0.45 km per year.

Table 4.4 shows changes in distance from the dam and elevation of the delta pivot point. It shows that the pivot point was almost stagnant in terms of its advancement, whereas a slight change in the elevation was observed. It might be due to keeping the minimum water level above the minimum pool level during these years. Figure 4.10 shows the water level during 2005, 2006 and 2007.

Table 4.4. Changes in pivot point of the sediment delta in the reservoir

Year	Elevation m+MSL	Distance from dam km
2003	418.71	12.75
2004	413.53	10.54
2005	413.41	10.54
2006	413.44	10.54
2007	413.75	10.54
2008	419.51	10.54

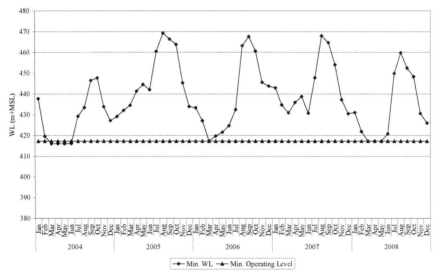

Figure 4.10. Minimum water levels in the reservoir during last five years

4.6.   Prospects of sediment management

For sediment management in the reservoir HR Wallingford (1998) suggested to focus on an operational policy which would keep the risks to the tunnel low, the construction of an underwater dyke to protect intakes and the construction of a low level flushing system, with a capacity of 7,000 m³/s, which would maintain quasi stable sedimentation conditions with a reservoir capacity of approximately 55% of the original value.

Later Haq and Abbas (2006) evaluated these suggestions for sediment management in the Tarbela Reservoir. They found that the sedimentation in the reservoir can be managed by means of a reservoir operational policy. Raising the minimum reservoir level every year by 1.2 m would result in deposition of sediments in the upper reaches of the reservoir and thus the advancement of the sediment delta can be delayed. Though this option entails no capital cost but would progressively result in increased loss of live storage. Minimum reservoir level of 417 m+MSL as fixed in 1998 is being maintained in order to use optimally the available storage.

Protection of tunnel intakes against sediment clogging by construction of an underwater dyke in front of the intakes as proposed by HR Wallingford was evaluated by Haq and Abbas (2006). They found that this option would not only involve tremendous stability and construction problems but also its benefits in the absence of sediment flushing from the reservoir seems minimal.

Reduction of sediment influx either by river basin management or by construction of check dams in the upper river basin is impractical as about 90% of total runoff is dominated by snow / glacier melt. Nothing can be done at this attitude on the steep mountains.

Annual evacuation of 200 MT of sediment by flushing through four low level high capacity outlets from the left bank was also proposed by the HR Wallingford. This option would comprise four 12 m diameter tunnels driven through the left abutment, possibly underneath the auxiliary spillway and discharging into its plunge pool. The abutment is weak. There have been a lot of problems and it has stabilized after a lot of remedial works. This proposal carries a large number of grey areas which need to be prudently addressed before taking it to a feasibility stage. WAPDA considers the underwater dyke and the four tunnels an unprecedented option, the example of which does not exist elsewhere in the World. Moreover, this option would in no time adversely affect the downstream hydropower Projects of Ghazi Barotha and Chashma and terminates them much earlier.

Measures in terms of dredging of sediments from this mega reservoir are almost impossible. The dredging option in case of Tarbela Reservoir is not only prohibitive in cost but is also without any precedence. Any dredging proposal to be effective must provide for removal and disposal of 500,000 tonnes of sediments every day. Realistically, the target is unattainable even if hundred of dredgers and ancillary equipment are deployed over the reservoir stretch of 50 km² to work round the clock.

Measures to increase the live storage capacity of the reservoir would entail raising of the crests of all embankment dams. Considering the existing foundation conditions at the site and other geotechnical problems of the embankment dams, this option poses serious stability threats to the dam. Therefore, this option is also discounted as being infeasible and impractical.

For flushing the delta should be close to the dam. The reservoir has to be depleted to its lowest level. Power house has to be closed. Discharges of the order of 5,600 m³/s would have to pass over the exposed delta, so that they can create shear velocity and entrain the deposited sediments. Large low level outlet capacity is required to pass the discharge. The outlets need to be steel lined to withstand the abrasion otherwise after flushing they would erode and it may not be possible to close the gates to refill the reservoir as happened in Volta dam. It may not be possible to refill the reservoir in a drought year. The reservoir is operated on irrigation demand and cannot be operated in flushing mode without the assurance of its refilling.

4.7.   Tarbela Dam and Pehure High Level Canal

PHLC takes off from the right bank of the Tarbela dam as shown in Figure 4.11. The sedimentation data from the Tarbela Reservoir and the studies cited above show that under prevailing conditions of sediment inflow and reservoir operation, the delta will keep on moving. Its rate of advancement would roughly be around 0.50 km per year. Now the delta is almost 10 km away from the dam and under these conditions it probably will take around 15-20 years in reaching close to the dam, which would be catastrophic for the outlets including PHLC.

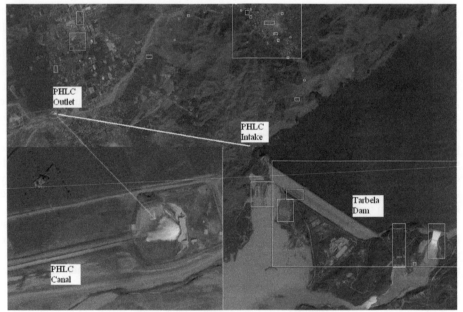

Figure 4.11. Tarbela Dam and PHLC

The intake levels of the PHLC and other tunnels are given in Figure 4.12 in comparison with the elevation of the pivot point in September, 2008. The intake level of PHLC is at an elevation of 393.2 m+MSL, which, under the current situation is about 26 metres below the pivot point elevation. If the delta continues to move, then along with other tunnels, PHLC would also undergo some severe sedimentation problems, which would ultimately cause operation and maintenance problems in PHLC. In the beginning, the PHLC would receive only the fine sediment with low concentrations but gradually, it would receive coarser sediment with even higher concentrations.

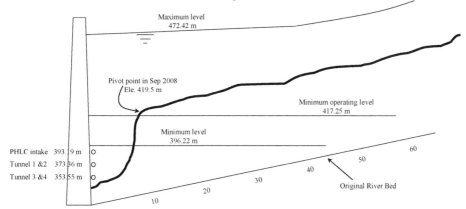

Figure 4.12. Schematics of the pivot point elevation in comparison to intakes invert levels

5. Hydrodynamic behaviour of Machai Maira and Pehure High Level Canals

5.1. Background

In demand based irrigation canals, water supplies to the canals depend upon crop water requirements (CWR) in the command area. The CWRs vary throughout the growing season. Hence flows in the canal also remain varying most of the time in order to keep a balance between supplies and demands. Frequent variations in flows have serious implication on canal's hydrodynamic performance, particularly in automatically controlled irrigation canals. In such circumstances if discharge control algorithms and canal operations are not properly planned then canal's safety and flow stability comes at stake. The unusual behaviour of any self regulating structure or any automation algorithm can cause troubles in the sustainability of the system.

Downstream control and Supervisory Control and Data Acquisition (SCADA) systems are usually employed in automatic demand based canal operations. Downstream control is a control technique in which the water level regulators respond to conditions in the downstream canal reach. The regulators in the main system maintain a constant water level at the downstream side of the structure, without regarding discharges. Such a regulation of structures means that more supply is given to a canal reach when the water level drops (Ankum, undated). The effect is that the discharge at each regulator is automatically adjusted to the accumulated downstream demand for irrigation.

The USC-PHLC system is a combination of fixed supply based and flexible demand based operations. The supply based system is fully manually controlled whereas the flexible demand based system is automatically controlled at main canal level and manually controlled at the secondary level. The question arises here how operation of manually controlled irrigation systems affects the hydrodynamics of the automatically controlled systems? For example how the automatic discharge controller responds to any change in the flow; how the amount of discharge variation (water used or refused) and the location of this change along the canal affects the stability and response times in the automatically controlled irrigation canals?

To see the effect of these operations on hydrodynamic behaviour of irrigation canals having upstream control with fixed supply based operation and the downstream control with flexible demand based operation, hydrodynamic modelling was performed. The Simulation of Irrigation Canals (SIC) Model was used to assess the effects of above mentioned canal operation scenarios. The model was calibrated with the field measurements before its application.

5.2.   Objectives of the hydrodynamic modelling

The overall aim of the hydrodynamic modelling study was to understand the hydrodynamic behaviour of the irrigation canals under study. For this purpose the objectives were formulated as to:

- study the hydrodynamic behaviour of Machai Branch Canal, Maira Branch Canal and PHLC (Figure 5.1.B) under steady state and unsteady state conditions;
- assess the effect of PI discharge controllers on the hydrodynamic performance of the canals under study;
- assess the effects of various options of Crop Based Irrigation Operations on the automatic hydraulic behaviour and stability of the canal.

5.3.   Methods and materials

5.3.1. Irrigation infrastructure

The irrigation system under study consists of three canals Machai Branch Canal, Maira Branch Canal and Pehure High Level Canal (PHLC). The Machai Branch canal is located at the upstream end and gets water from Swat River through Upper Swat Canal and the PHLC gets water from Indus River through Tarbela Reservoir. PHLC falls into Machai Branch Canal at a confluence downstream of RD 242. The RD 242 is a division point between upstream control and downstream control, at an abscissa of 74,000 m at Machai Branch Canal.

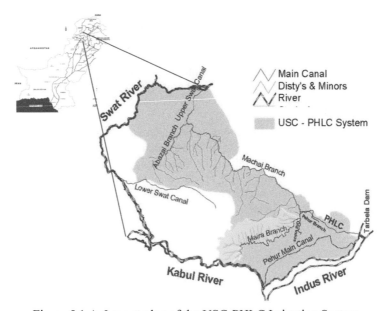

Figure 5.1.A. Layout plan of the USC-PHLC Irrigation System

The Machai Branch Canal upstream of RD 242 has manual supply based flow regulation, whereas downstream of RD 242 it has automatic downstream control flow regulation. These canals feed their own secondary system as well as supply water to the Maira Branch Canal. The design contributions to Maira Branch Canal from Machai Branch Canal is 14 m^3/s and from PHLC 24 m^3/s. These supplies to Maira Branch Canal are variable as it has demand based operations. The maximum discharge capacity at the confluence is 32 m^3/s. At confluence point, the PHLC joins Machai Branch Canal, which is located at 300 m downstream of RD 242. The plan of the area under study, the schematic layout of the irrigation canals as well as their flow directions given in Figures 5.1.A, 5.1.B and 5.1.C. The salient features of the main canals are given in Table 5.1.A, whereas the design discharges of secondary offtakes are given in Table 5.1.B.

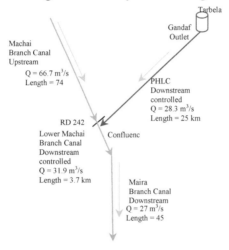

Figure 5.1.B. Flow directions in Machai, Maira and PHLC irrigation canals

Table 5.1.A. Salient features of irrigation canals under study

S. No.	Description	Machai Branch Canal	Pehure High Level Canal	Lower Machai (d.s. RD 242)	Maira Branch Canal
1	Discharge (m^3/s)	66.7	28.3	31.9	27.0
2	Culturable Command Area (ha)	48,556	5,100	6,728	29,000
3	Length	73.8	25.5	3.74	44.8
4	Cross-section	Trapezoidal (unlined)	Parabolic Lined	Trapezoidal (sides lined)	Trapezoidal (unlined)
5	Cross regulators	12 No. Radial gates	03 No. AVIO, 02 No. AVIS		08 No. AVIS gates
6	Discharge and water level regulation	Manual	Automatic	Automatic	Automatic
7	Velocity (m/s)	0.85-1.04	1.39	0.59	0.59
8	Bed slope	0.0002	0.00018	0.00018	0.00018
9	Offtakes (secondary)	21	5	3	17
11	Direct outlets	69	23	10	62

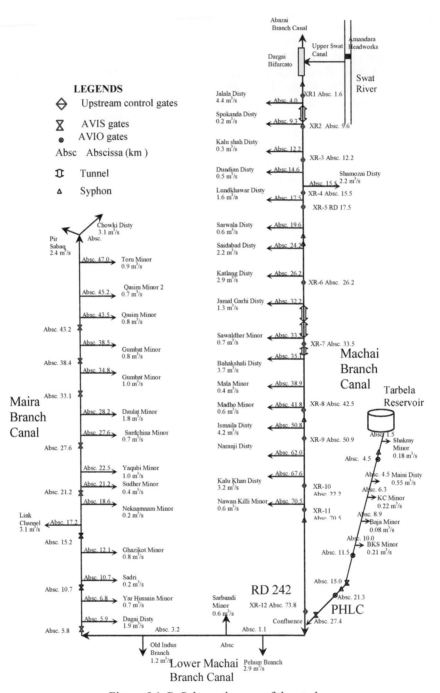

Figure 5.1.C. Schematic map of the study area

Table 5.1.B. Design discharges of secondary offtakes

Abscissa (m)	Parent canal	Offtake	Design Q (m³/s)
3993	Machai Branch Canal	Jalala Disty	4.4
9280	Machai Branch Canal	Spokanda Disty	2.3
12227	Machai Branch Canal	Kalu Shah Disty	0.2
14657	Machai Branch Canal	Dundian Disty	0.5
15483	Machai Branch Canal	Shamozai Disty	2.2
17538	Machai Branch Canal	Lund Khawar Disty	1.6
19556	Machai Branch Canal	Sarwala Disty	0.6
24152	Machai Branch Canal	Saidabad Disty	2.2
26187	Machai Branch Canal	Katlang Disty	2.9
32227	Machai Branch Canal	Jamal Garhi Disty	1.3
33289	Machai Branch Canal	Sawal Dher Minor	0.7
35128	Machai Branch Canal	Bhakshali Disty	3.7
38877	Machai Branch Canal	Mala Minor	0.4
41760	Machai Branch Canal	Madho Minor	0.6
50808	Machai Branch Canal	Ismaila Disty	4.3
62033	Machai Branch Canal	Naranji Minor	1.3
67569	Machai Branch Canal	Kalu Khan Disty	3.2
70518	Machai Branch Canal	Nawan Killi Minor	0.5
1527	PHLC	Shakray Minr	0.2
4478	PHLC	Maini Disty	0.5
6296	PHLC	Kotha College Minor	0.2
8864	PHLC	Baja Minor	0.1
9952	PHLC	Bamkhel South Minor	0.2
	Downstream RD 242		
1425	Lower Machai Branch Canal	Pehur Branch	2.9
1870	Lower Machai Branch Canal	Sarbandi Minor	0.6
3470	Lower Machai Branch Canal	Old Indus Branch	1.2
5910	Maira Branch Canal	Dagai Disty	1.9
6800	Maira Branch Canal	Yar Hussain Minor	0.7
10740	Maira Branch Canal	Sadri Minor	0.2
12110	Maira Branch Canal	Ghazi Kot Minor	0.8
17170	Maira Branch Canal	Link Cahnnel	3.1
18595	Maira Branch Canal	Nek Nam Minor	0.2
21230	Maira Branch Canal	Sudher Minor	0.4
22490	Maira Branch Canal	Yaqubi Minor	1.0
27630	Maira Branch Canal	Sard China Minor	0.7
28160	Maira Branch Canal	Daulat Minor	1.8
34760	Maira Branch Canal	Gumbat-1 Minor	1.0
38490	Maira Branch Canal	Gumbat - 2 Minor	0.8
43530	Maira Branch Canal	Qasim-1 Minor	0.8
45210	Maira Branch Canal	Qasim-2 Minor	0.7
46980	Maira Branch Canal	Toru Minor	0.9
48020	Maira Branch Canal	Pir Sabaq Disty	2.4
48020	Maira Branch Canal	Chauki Disty	3.1

5.3.2. System operations

The system has two modes of operation, fixed Supply Based Operation (SBO) in Machai Branch Canal and Crop Based Irrigation Operations (CBIO) in Maira Branch Canal and partly in PHLC. In SBO the canals are operated always at full supply discharge depending upon the water availability at the source. In CBIO the canals are operated according to the crop water requirement. CBIO will be further described in the subsequent section.

5.3.3. Crop Based Irrigation Operations (CBIO)

Crop Based Irrigation Operations (CBIO) is a canal operation strategy in which irrigation water supplies are made compatible with the command area crop water requirements (CWR). The CWRs are low in the beginning and end of the cropping season and high in the middle. Hence in CBIO the same trend is followed for supplying water to irrigation canals. Less water is supplied during low requirements and maximum water is supplied during peak requirements. Lower Machai (downstream of RD 242) and Maira Branch canal systems are operated according to CBIO. When the supplies fall below 80% of the full supply discharge, a rotation system is introduced amongst the secondary offtakes. During very low crop water requirement periods the supplies are not reduced beyond a minimum of 50% of the full supply discharge. These are the operational rules of CBIO, which were envisaged during system design in order to avoid sedimentation in the irrigation canals (Pongput, 1998). CBIO schedule for the year 2006-2007 is shown in the Figure 5.2.

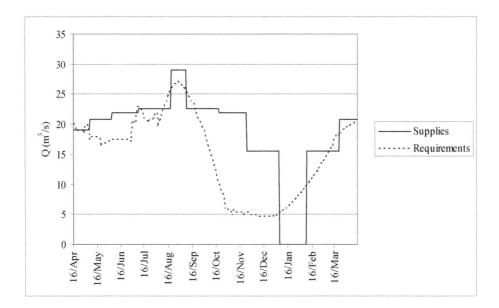

Figure 5.2. Flow requirements of area downstream of RD 242 and supplies decided under CBIO during 2006-2007

5.3.4. Discharge and water level controls

Supplies at the Machai Branch head and water levels in the canal are controlled by manually operated sluice gates. In PHLC the supplies at the head are controlled automatically at Gandaf Tunnel Outlet (GO) by automatic discharge and water level controllers. The Gandaf Tunnel Outlet has been equipped with Supervisory Control and Data Acquisition (SCADA) system for automatic discharge control, having a PI discharge control algorithm. The water levels in the canal are also controlled automatically by downstream control hydro-mechanically operated AVIS and AVIO type gates. Similarly water levels in Maira Branch canal are controlled by AVIS type cross regulators.

5.3.5. Secondary offtakes operation

The operation of the secondary system is manual in both Machai Branch Canal and Maira Branch Canal and is automatic in PHLC except Maini Distributary. The discharge regulating structures on supply based and demand based secondary offtakes are shown in Figure 5.3.A and 5.3.B. In Machai Branch Canal the secondary offtakes are operated according to the water availability in the canal and in Maira Branch the secondary offtakes are operated according to the crop water requirements. The water use or refuse in Machai Branch Canal depends upon water availability at the source whereas in Maira Branch Canal and PHLC it depends upon crop water demands in their command areas.

A. Manually controlled offtake   B. Float chamber in automatically
                                          controlled offtake

Figure 5.3. Head regulator in manually controlled Yar Hussain Minor and float chamber in automatically controlled Baja Minor.

5.3.6. Modelling canal operations

The assessment of hydrodynamic behaviour of an irrigation canal network under varying flow conditions is a prerequisite for attaining efficient system operations. The computer model Simulation of Irrigation Canals (SIC) has been used in this study to assess the hydrodynamics of the irrigation canals under study. The SIC model will b e described in detail in Chapter 7.

5.3.7. Calibration of the model

The model was calibrated in steady state conditions by measuring water levels and discharges in the Machai Branch Canal, Maira Branch Canal and in the PHLC. For Maira Branch Canal and PHLC calibration the canals were kept running for two to three days to obtain steady state conditions. Then these canals were divided into three parts for the discharge and corresponding water level measurements. All of the inflows and outflows to and from these parts were measured. The resulting water levels were measured at the upstream and downstream side of every cross regulator. The measurement campaigns in Maira Branch Canal and PHLC were conducted in August 2005 and November 2005 respectively. In Machai Branch Canal the Manning's roughness value was determined directly at selected reaches in July 2007. Some suitable straight long sections for 1 to 2 km were selected at head middle and tail of Machai Branch Canal. These sections had no or least backwater effect. These sections were also used for sediment sampling and their exact locations have been given in Figure 6.1. The measured values for Machai Branch Canal were used directly for modelling, whereas for Maira Branch Canal and PHLC the simulated water levels were compared with the measured water levels and the canal roughness value was adjusted. For validation purposes the field data measured in July 2007 was used for Maira Branch Canal. There was no measurement of discharge in PHLC after November 2005 so PHLC calibration results were not validated.

The following criteria proposed by Jabro et al. (1998) were used for model calibration:

$$ME = \max |M_i - S_i|_{i=1}^{n} \tag{5.1}$$

$$RMSE = \left[\frac{1}{n} \sum_{i=1}^{n} (M_i - S_i)^2 \right]^{0.5} \times \left(\frac{100}{M} \right) \tag{5.2}$$

$$EF = \frac{\sum_{i=1}^{n} (M_i - M)^2 - \sum_{i=1}^{n} (M_i - S_i)^2}{\sum_{i=1}^{n} (M_i - M)^2} \tag{5.3}$$

$$CRM = \frac{\sum_{i=1}^{n} M_i - \sum_{i=1}^{n} S_i}{\sum_{i=1}^{n} M_i} \tag{5.4}$$

$$MAE = \frac{1}{n} \sum_{i=1}^{n} |M_i - S_i| \qquad (5.5)$$

Where M_i, S_i and M are the measured, simulated and average of measured values. The maximum error *(ME)* is a measure of the maximum error between any pair of simulated and measured values. The lower limit and the best value of *ME* is zero. The root mean square error *(RMSE)* provides a percentage for the total difference between simulated and measured values proportionated against the mean observed values. The lower limit for *RMSE* is zero and indicates a more accurate simulation. The modelling efficiency *(EF)* is a measure for assessing the accuracy of simulations. The maximum value for *EF* is one, which occurs when the simulated values match the measured values perfectly. The coefficient of residual mass *(CRM)* is an indication of the consistent errors in the distribution of all simulated values across all measurements with no consideration of the order of the measurements. A CRM value of zero indicates no bias in the distribution of simulated values with respect to measured values. The mean absolute error *(MAE)* is the mean error estimation, which is better when closer to zero.

5.4. Results and discussion

5.4.1. Model calibration and validation

The model was run in steady state conditions according to the field conditions. The discharge withdrawn by the offtakes was imposed on these offtakes in the simulation. The Manning's roughness values were adjusted for Maira Branch Canal and for PHLC. The calibrated values of Manning's roughness were quite close to the design values in Maira Branch Canal and PHLC. It might be due to the recent remodelling of the Maira Branch Canal and the recent construction of PHLC. The Manning's roughness values for Maira Branch Canal vary from 0.020 to 0.024 at different reaches, whereas for PHLC these values range from 0.014 to 0.017. Table 5.2 presents the parameters used for model calibration and validation. High values of EF and low values of ME, CRM, RMSE and MAE during calibration and validation for Maira Branch Canal and for PHLC show that the model can be used for further simulations.

Table 5.2. Results of model calibration and validation for water flow of Maira Branch Canal and for PHLC

S. No	Parameter description	Maira Branch Canal		PHLC
		Calibration	Validation	Calibration
1	ME (m)	0.276	0.294	0.061
2	RMSE	0.169	0.187	0.402
3	EF	0.996	0.991	0.998
4	CRM	-0.001	-0.012	-0.002
5	MAE (%)	-0.116	-0.128	0.214

5.4.2. Steady state simulations

A number of different scenarios were investigated in order to study the canal's hydrodynamic behaviour under steady state conditions. These scenarios will be discussed in detail in the following.

Scenario-1. Design conditions

It is generally desirable to see how the canal behaves under full supply discharge or design discharge. Therefore first simulations were made to test the canal's hydraulic behaviour under design conditions. In these simulations it has been tested whether under the design hydraulic features the canals can deliver and distribute the target discharges to the secondary system. Table 5.3 gives information on the discharges supplied at the Machai Branch Canal and PHLC heads and the Figures 5.4.A, 5.4.B and 5.4.C present the resulting water surface profiles.

Table 5.3. Discharges supplied under design conditions

S. No	Canal	Q supplied (m³/s)	Design losses	Remarks
1	Machai Branch Canal	66.7	6%	Design discharge
2	PHLC	27.1	3%	Excluding Jandaboka Lift Scheme
3	Lower Machai Branch Canal at Confluence	32.1	6%	Calculated at the confluence from the two sources, Machai Branch Canal and PHLC

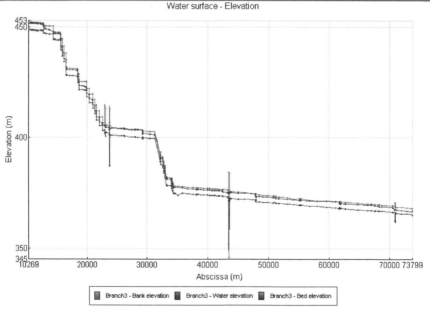

Figure 5.4.A. Water surface profile in Machai Branch Canal under design discharge.

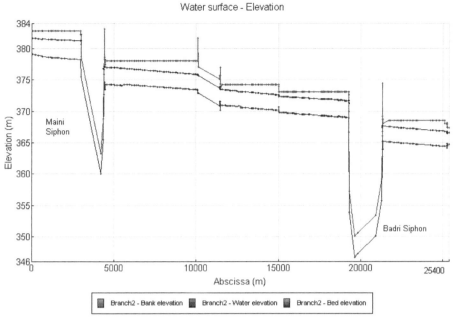

Figure 5.4.B. Water surface profile in PHLC under design discharge

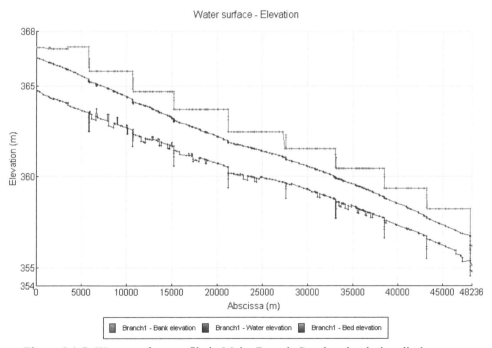

Figure 5.4.C. Water surface profile in Maira Branch Canal under design discharge.

Under these conditions all the offtakes got their design discharge except Pehure Branch, Old Indus Branch and the Link Channel of Maira Branch Canal. These offtakes could not get the design discharge because of low water level in the canal. The discharge supplies to these offtakes are given in Table 5.4. An amount of 22 m³/s discharge was escaped at the Machai tail upstream of RD 242 because it was more than the target supplies in the system downstream of RD 242.

Table 5.4. List of offtakes not getting water under design conditions

S. No	Offtake	Q supplied (m³/s)	Q design (m³/s)	% difference	Water level in main canal (m+MSL)
1	Pehure Branch	2.6	2.9	10	366.29
2	Old Indus	1.1	1.2	7	365.90
3	Link Channel	2.3	3.1	24	362.76

The water levels under design conditions upstream and downstream of Machai Branch Canal and Maira Branch Canal cross regulators are given in Tables 5.5.A, 5.5.B and 5.5.C, where SF stands for submerged flow and FF stands for free flow condition.

Table 5.5.A. Water levels upstream and downstream of cross regulators of Machai Branch Canal

Cross regulator	Abscissa (m)	Q (m³/s)	u/s WL (m+MSL)	d/s WL (m+MSL)	Gate opening (m)	Flow condition
Machai XR-3	12,258	58.9	451.31	451.04	3.00	SF
Machai XR-4	15,514	55.8	446.79	445.84	3.00	FF
Machai XR-5	18,334	54.0	430.37	428.71	2.50	FF
Machai XR-6	26,226	47.8	403.74	403.42	2.50	SF
Machai XR-7	35,840	41.4	377.19	376.86	2.50	SF
Machai XR-8	42,492	39.9	375.64	375.21	2.50	SF
Machai XR-9	51,208	35.1	372.63	372.23	2.50	SF
Machai XR-10	67,577	31.0	369.04	368.75	2.50	SF
Machai XR-11	70,552	30.0	368.12	367.60	2.00	FF
Machai XR-12	73,762	7.9	366.55	366.52	2.00	SF

Table 5.5.B. Water levels upstream and downstream of cross regulators of PHLC

Cross regulator	Abscissa (m)	Q (m³/s)	u/s WL (m+MSL)	d/s WL (m+MSL)	Gate Opening (m)	Flow condition
PHLC XR-1	4,442	26.7	377.21	377.10	2.25	SF
PHLC XR-2	11,464	25.2	373.84	373.52	1.58	SF
PHLC XR-3	15,011	24.7	372.49	372.34	1.57	SF
PHLC XR-4	21,301	24.3	367.75	367.66	2.25	SF
PHLC XR-5	25,266	24.2	366.65	366.51	1.57	SF

Table 5.5.C. Water levels upstream and downstream of cross regulators of Maira Branch Canal

Cross regulator	Abscissa	Q	u/s WL	d/s WL	Gate Opening	Flow Condition
	(m)	(m³/s)	(m+MSL)	(m+MSL)	(m)	
Maira XR-6	5,821	27.0	365.27	365.16	1.76	SF
Maira XR-7	10,682	24.0	364.10	363.96	1.57	SF
Maira XR-8	15,200	21.0	363.10	362.99	1.57	SF
Maira XR-9	21,170	17.2	361.91	361.84	1.57	SF
Maira XR-10	27,570	13.6	360.98	360.91	1.40	SF
Maira XR-11	33,070	10.9	359.95	359.89	1.20	SF
Maira XR-12	38,420	9.4	358.86	358.80	1.02	SF
Maira XR-13	43,170	8.2	357.78	357.71	0.80	SF

Scenario-2. Machai system requirements plus its minimum contribution to Maira

As Machai Branch Canal contributes to Maira Branch Canal as well as serving its own secondary system. In case of water shortage in Swat River, the priority is given to Machai Branch Canal's secondary system for water deliveries. Therefore this scenario has been simulated with maximum deliveries to Machai Branch's system and minim contribution to Maira Branch Canal. A discharge of 46.5 m³/s has been released from the Machai Branch Canal headworks and its effect on the water delivery to Machai offtakes has been assessed. It was found that all the Machai offtakes upstream of RD 242 achieved their design discharges except Jamal Garhi Disty and Nawan Killi Minor. The water levels upstream and downstream of Machai Branch Canal cross regulators have been given in Table 5.6.A and the gate openings of the offtakes have been given to satisfy the design discharges in Table 5.6.B.

Table 5.6.A. Water levels (m) upstream and downstream of cross regulators

Cross regulator	Abscissa	u/s WL	d/s WL	Gate Opening	Flow Conditions
	m	m+MSL	m+MSL	m	
Machai XR-3	12,258	450.80	450.55	3.00	FF
Machai XR-4	15,514	446.49	445.69	2.50	FF
Machai XR-5	18,334	430.01	428.57	2.50	SF
Machai XR-6	26,226	403.18	402.87	2.50	SF
Machai XR-7	35,840	376.73	376.46	2.50	FF
Machai XR-8	42,492	375.24	374.81	2.50	SF
Machai XR-9	51,208	372.05	371.74	2.50	SF
Machai XR-10	67,577	368.44	368.24	2.50	SF
Machai XR-11	70,552	367.58	367.29	2.00	SF
Machai XR-12	73,762	366.57	366.49	2.00	SF

In order to supply the design discharges to Jamal Garhi Disty and Nawan Killi Minor the cross regulators (XR) No. 7 and XR No. 11 need to be operated respectively. The XR-7 was lowered down but no effect was observed on Jamal Garhi Distributary because of a tunnel in between the cross regulator and the Jamal Garhi Disty's head regulator. The Nawan Killi Mino was managed to get the design discharge by lowering the gates of XR-11 to an opening of 0.75 m.

Table 5.6. B. Offtake gate openings to satisfy the design discharges under scenario-2

S. No.	Offtake	Gate opening required (m)	Gate opening maximum (m)	Design Q (m^3/s)	Remarks
1	Kalushah Disty	0.10	0.91	2.9	
2	Dundian Disty	0.23	0.61	0.5	
3	Shamozai Disty	0.16	1.22	2.3	
4	Lundkhawar Disty	0.40	0.61	1.6	
5	Sarwala Disty	0.43	0.61	0.6	
6	Saidabad Disty	0.47	0.91	2.2	
7	Katlang Disty	0.57	0.91	2.9	
8	Jamal Garhi Disty	0.91	0.91	1.3	Q not achieved by 29%
9	Sawal Dher Minor	0.23	0.61	0.7	
10	Bakshali Disty	0.36	1.22	3.6	
11	Mala Minor	0.24	0.61	0.4	
12	Madho Minor	0.43	0.61	0.6	
13	Ismaila Disty	0.69	0.91	4.2	
14	Kalu Khan Disty	0.47	0.91	3.2	
15	Nawankilli Minor	0.61	0.61	0.6	Q not achieved by 19%

Scenario-3. Minimum discharge from Machai Branch Canal

The water levels under minimum discharge conditions at Machai Branch Canal cross regulators are given in Table 5.7.

Table 5.7. Water levels at cross regulators of Machai Branch Canal under minimum contribution from the Machai Branch Canal to Maira Branch Canal

Cross Regulator	Abscissa (m)	Q (m^3/s)	u/s WL (m+MSL)	d/s WL (m+MSL)	Gate opening (m)	Flow condition
Machai XR-3	12,258	37.1	450.43	450.18	3.00	SF
Machai XR-4	15,514	34.0	446.26	445.58	3.00	FF
Machai XR-5	18,334	32.1	429.74	428.46	2.50	FF
Machai XR-6	26,226	26.0	402.75	402.45	2.50	SF
Machai XR-7	35,840	19.6	376.35	376.14	2.50	SF
Machai XR-8	42,492	18.1	375.05	374.54	0.65	FF
Machai XR-9	51,208	13.3	372.07	371.33	0.50	FF
Machai XR-10	67,577	9.2	368.61	368.01	0.35	SF
Machai XR-11	70,552	8.2	367.83	366.10	0.30	FF
Machai XR-12	73,762	7.8	366.54	366.51	2.00	SF

As in Machai Branch Canal the discharge availability predominantly depends upon water availability in the Swat River and it gets usually less water in winter especially in November and December. This scenario has been simulated to assess the effect of low flows in Machai Branch Canal on the water delivery and distribution. An amount of 37 m^3/s water was supplied at Machai Branch Canal headworks, whereas the rest of the discharge supplies remained same as in Table 5.3. It was observed that the secondary offtakes given in Table 5.8 could not get the design discharge.

Table 5.8. Offtakes of Machai Branch Canal deprived of design discharge with low supplies in Machai Branch Canal

S. No.	Offtake	Abscissa m	Q supplied m^3/s	Q design m^3/s	Percent difference %	WL in main canal m+MSL
1	Jamal Garhi Disty	32,229	0.6	1.3	52	390.08
2	Madho Minor	41,762	0.5	0.6	11	375.03
3	Ismaila Disty	50,810	1.9	4.2	55	371.68
4	Kalu Khan Disty	67,572	1.5	3.2	52	367.92
5	Nawan Killi Minor	70,512	0.1	0.6	87	367.15

These offtakes can be supplied with the design discharge with cross regulators adjustment. In Table 5.9 the required adjustment of the cross regulators has been given for achieving the design discharges. There is a tunnel in between the Jamal Garhi Disty and the Machai Branch Canal cross regulator No. 7 at abscissa 35,840. It was found that there was hardly any effect of cross regulator operation on the discharge of Jamal Garhi Disty. So Jamal Garhi Disty could not get its design discharge. Whereas the remaining of the deficient offtakes achieved their design discharges.

Table 5.9. Cross regulators opening for getting design discharge through the discharge deficient offtakes

S. No.	Cross regulator	Gate opening (m)	Offtakes supported	New WL in the canal (m)	Raise in WL (m)
1	Machai XR-8	0.65	Madho Minor	375.13	0.10
2	Machai XR-9	0.50	Ismaila Disty	372.09	0.42
3	Machai XR-10	0.35	Kalu Khan Disty	368.31	0.39
4	Machai XR-11	0.30	Nawankilli Minor	367.83	0.32

5.4.3. Performance of Proportional Integral discharge controllers

Evaluation of PI coefficients

PI coefficients play key role in water regulation in the continuous automatic flow control systems. The selection of correct values of the proportional, integral and derivative gain factors leads to safe and stable operations of the canal and prevents any oscillatory behaviour of the automatically regulated hydro-mechanical gates. The quick response of the discharge regulator to the deviations from the setpoint (proportional property) and meeting the setpoint (the integral property) are the characteristics of PI discharge controllers. Various values of these coefficients have been tested by trial and error and some optimum values were found which would lead to desired system operations.

Kp (Proportional gain) coefficient

Three different values of proportional gain coefficients have been tested as given in Table 5.10 along with other information on the refusal of the discharge and the locations. In Figure 5.5 the results of the simulations have been presented where the target water level is 382.15 m+MSL.

Table 5.10. Different values of Kp and other parameters

S. No.	K_p value	Q from Machai Branch (m^3/s)	Q at Gandaf Outlet (m^3/s)	Q at confluence (m^3/s)	Q refused (% of Q at confluence)	Location of closed offtakes
1	1.30	8.0	17.5	25.5	24%	Head reach
2	2.00	8.0	17.5	25.5	24%	Head reach
3	2.50	8.0	17.5	25.5	24%	Head reach

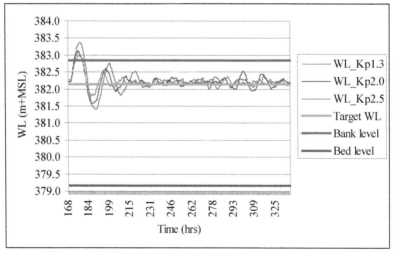

Figure 5.5. Water level oscillations under different K_p values

An amount of 6.1 m³/s discharge was refused at the head of the Maria Branch Canal by closing five secondary offtakes. The discharge released under three different values of K_p factor, as given in Table 5.10, was tested and its effect was observed on the pattern of flow releases from the Gandaf Outlet (GO). The results are given in Figure 5.5, where WL means actual water level at different K_p values and Target WL means required water level at canal head reach.

Figure 5.6 shows that the K_p = 1.30, which is basically used at the Gandaf Tunnel Outlet gives oscillatory behaviour and requires long time to reach at new steady state conditions with a steady state error or 0.05 m, whereas the K_p values 2.0 and 2.50 give comparatively less oscillations and the discharge gets stable earlier. The discharge released against these K_p values is given in Figure 5.6, which also shows almost the same behaviour. The discharge released under K_p = 1.30 gets stable after about 64 hours, whereas it gets stable under K_p = 2.00 and K_p = 2.50 after 46 hours. The maximum and minimum discharge released under all these K_p values is almost the same.

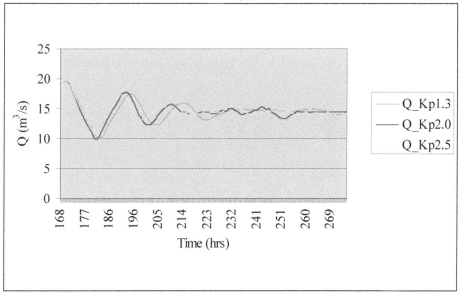

Figure 5.6. Discharge released under different K_p values

Testing of Integral time (Ti) coefficient

The integral time, T_i, is another important coefficient which affects the response of the PI controller. The integral property reduces the decrement and brings the derivations to zero. Two values of integral time, T_i = 3,000 and 1,200 seconds were simulated with the same amount of discharge variation. It has been tested which value brings the stability earlier in the system. The results are presented in Figure 5.7. The flow parameters during this test were the same as given in Table 5.10 with K_p value of 2.5. Figure 5.7 shows that T_i = 1,200 seconds leads to more oscillations as compared to the T_i = 3,000. As far as stability is concerned, the discharge gets stable earlier under T_i = 3,000.

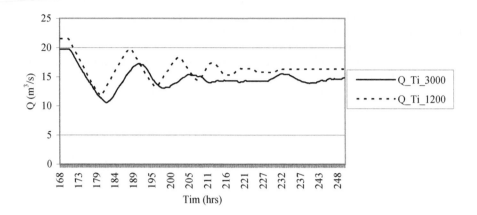

Figure 5.7. Discharge variations under T_i = 3,000s and 1,200s

5.4.4. Effect of amount and location of discharge refusal on discharge control

Effect of location of discharge refusal

In this scenario the effect of discharge refusal from different locations along the canal has been simulated to assess the response times and system stability against water refusal at various distances. The effect of the location of offtakes closed on the system stability has been compared and results are presented in Figure 5.8. The offtakes were grouped with almost the same amount of discharge at the head portion, tail portion and all along the canal (mixed). Their effects were simulated on the system behaviour. Table 5.11 gives information about offtakes grouping, their location and the groups' total discharge.

Table 5.11. Information on the offtakes grouping

Group no.	Location	Total discharge refused (m³/s)	Percentage of flow at confluence	Offtakes names
1	Head	6.1	24%	Pehure Branch, Sarbandi Minor, Old Indus Branch, Dagi Disty, Yar Hussain Minor
2	Tail	6.1	24%	Gumbat II Minor, Qasim I & II Minors, Toru Minor, PirSabaq Disty
3	Mixed	5.2	21%	Pehure Branch, Dagi Disty, Yaqubi Minor, Gumbat II Minor

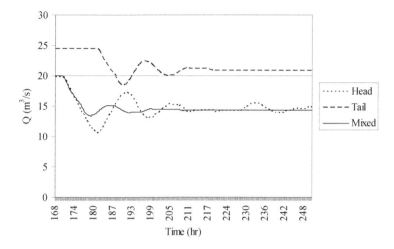

Figure 5.8. Flow stability and response times of offtakes closure at different locations

Figure 5.8 shows that the composite grouping of offtakes gives early stability and has much less oscillations as compared to the discharge refusal at head and tail. The discharge controller responds to discharge variations at the tail after 14 hours and gets stable after 51 hours. Whereas the discharge controller responds after 3 hours to discharge refusal at head and gets stable after 51 hours. The discharge controller reacts after 3 hours to the discharge refusal along the canal (composite) and gets stable after 29 hours. These results show that the mixed offtakes closing is a better option for stable system operations.

Effect of amount of discharge refusal

The amount of discharge refusal also affects the stability of the system and response times. The effects of two different amounts of discharge were compared and the results are presented in Figure 5.9, which shows that high amounts of discharge refusal take more time for system stability, whereas in the less discharge refusal situations the system stabilizes comparatively earlier. For the 50% discharge refusal the system took 48 hours for stability whereas in case of discharge refusal of 24% the system got stable in 28 hours.

5.4.5. Testing of CBIO schedules

The overall purpose of this study is to describe the hydrodynamic behaviour of automatically downstream controlled systems under the CBIO. To assess the hydrodynamic behaviour of the canals and the system stability under these operations is very important for efficient and reliable system operations. Hence four different discharges of CBIO were simulated and the results are presented in Figure 5.10. The possible CBIO discharges were tested by changing flow demands as 100%, 80%, 67%, 50% of the design supply and then again on 100%.

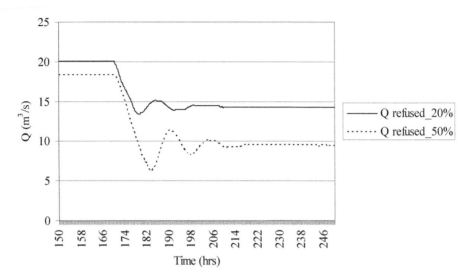

Figure 5.9. Response times of 20% and 50% offtakes closure along the canal

Figure 5.10 shows that the gradual increase or decrease in flow conditions gets stability earlier and takes less response time, whereas the big changes in discharge refusal or discharge opening result in prolonged instability and longer response times as given in Table 5.12. Table 5.12 shows that as the amount of discharge variation increases or decreases the response time increases and decreases accordingly.

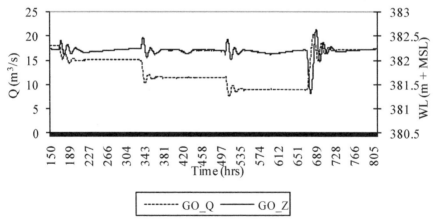

Figure 5.10. Testing of CBIO schedule at 100, 80, 67, 50 and 100% of full supply discharge

Table 5.12. Response times of different flow changes

S. No	Flow variations (%)	Amount of flow supplies at confluence (m³/s)	Response time (hrs)
1	100 → 80	2.9	32.83
2	80 → 67	3.6	44.33
3	67 → 50	2.5	38.67
4	50 → 100	8.2	64.17

5.4.6. Gate responses

The AVIS/AVIO gates are self operated gates and attain new positions automatically by hydro-mechanical action after any change in discharge or water level in the canal. Therefore gate responses to discharge and water level variations are crucial for smooth and secure irrigation canal operations. It needs to be assured that the variation in water withdrawal or refusal in the canal may not lead to abrupt gate opening/closing or oscillations. Hence the gate behaviour under some discharge refusals has been tested and is presented in Figure 5.11. A discharge of 6.0 m³/s was refused at the tail portion of the Maira Branch Canal. The discharge refusing point was selected at the tail portion so that the behaviour of all of the automatic cross regulators can be assessed. Figure 5.11 shows that the gates settled smoothly to the new positions within 3-6 hours.

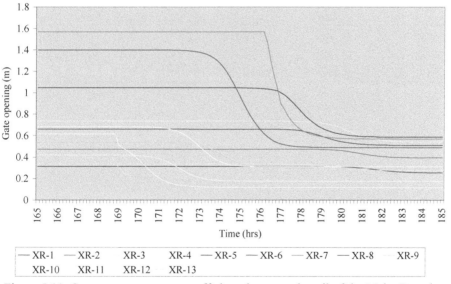

Figure 5.11. Gate responses to some offtakes closure at the tail of the Maira Branch Canal

Together with the smooth settling and opening of the gates, it is also important to know the reaction time for estimating the time elapsed in travelling of the effect any disturbance in the canal to the controller. In this case the disturbance was created at the tail of the Maira Branch Canal, whereas the controllers were at the head of PHLC. Table 5.13 gives the reaction times of the cross regulators from cross regulator No. 13 (XR-13) at Maira Branch Tail to cross regulator No. 1 (XR-1) at PHLC head, which finally conveys the messages of change to water level sensor and discharge control system. The total time elapsed from XR-13 to XR-1 was 10.67 hours and the final settlement of XR-1 took place after 16.67 hours. Every next cross regulator took about 0.89 hours to respond and finally it settled on the new position after 5.58 hours on average.

Table 5.13. Gate openings and time elapsed in response of discharge refusal at Maira tail.

Cross-regulator	Abscissa from PHLC Head (m)	Start time (hrs)	End time (hrs)	Initial opening (m)	Final opening (m)
XR-13	67,124	168.33	171.83	0.40	0.11
XR-12	62,374	168.83	173.67	0.67	0.12
XR-11	57,024	169.33	175.00	0.42	0.18
XR-10	51,524	170.33	175.67	0.68	0.31
XR-9	45,124	171.33	177.17	0.54	0.31
XR-8	38,794	172.00	178.50	1.40	0.49
XR-7	34,674	176.17	180.50	1.60	0.57
XR-6	29,774	174.83	183.00	1.05	0.59
XR-5	2,4254	176.50	182.33	0.66	0.51
XR-4	21,312	176.83	182.33	0.34	0.30
XR-3	15,032	177.67	183.67	0.73	0.54
XR-2	11,485	178.33	183.33	0.47	0.40
XR-1	4,487	179.00	185.00	0.32	0.25

6. Sediment transport in Machai and Maira Branch Canals

In order to assess sediment transport in the canals under study, water and sediment measurements were performed for almost one and half year from January 2007 to July 2008. Operation data of the main canal and secondary canals was collected from the Department of Irrigation, Khyber Pakhtunkhwa for two years (four cropping seasons) from October 2006 to September 2008 to determine the water supply to the main canals and then water distribution to the secondary canals. These data sets provide complete information on the water and sediment balances in the canals under study during the study period.

Three sets of measurements were conducted in July 2007, December 2007 and then in July 2008 to study the water and sediment mass balances under different conditions. These measurements were conducted keeping in view the different flow and sediment transport conditions. The July measurements represent the monsoon and flood season, so more water and sediment is expected in the system, whereas December measurements are in the winter season, so less water and sediment are generally available in the system. The water distribution in the canal in these seasons is different. In addition the sediment inflow to the irrigation canals was monitored on weekly basis (during peak sediment inflow season) and monthly basis (during low sediment inflow season). The system was divided into two parts upstream of RD 242 and downstream of RD 242.

Main purpose of field measurements was to test some management options in the irrigation canals, particularly in the downstream control part of the canals in order to manage sediment. However, it was difficult to convince the water users of the area and the Department of Irrigation to change the canal operation pattern for sediment management. The reason behind this was normal sediment inflow to the irrigation canals under study during the study period. There was not a severe sedimentation problem in these irrigation canals, as is expected after the sediment discharge from the Tarbela Reservoir in near future. Therefore the measurements were conducted on usual canal operation, under different conditions, which can be used to develop and calibrate some sediment transport model. Then various scenarios water and sediment inflows with various water and sediment management options can be simulated.

A total number of three mass balance studies were conducted. In these mass balance studies the detailed hydraulic and sediment transport functions of the canal were determined and the interdependence between hydraulic, morphologic and sediment parameters was found. The parameters like water discharge, canal cross-section, suspended sediment load, bed material and water surface slope were measured during the mass balance studies. The canal from Machai Branch Canal's head to the tail of Maira Branch Canal was divided into seven reaches. All of the water and sediment inflows and outflows to and from these canal reaches were measured. The water and sediment (suspended load + bed material load) transport measurements were done at seven sampling points in the canal in order to know the distribution of the sediment along the canal. The outgoing water and sediment measurements were also made at all of the secondary offtakes. For the offtaking canals, the discharge was computed from the gauge reading (which were already calibrated), whereas the sediment was collected by taking boil samples immediately downstream of offtakes' head regulators.

In order to assess morphological changes and scouring/deposition in the canal, the cross-sectional and longitudinal surveys were conducted in the canal at four different time periods, in January 2007, July 2007, January 2008 and then in July 2008. In case of flowing water the bed levels were measured at thirty to forty locations along the canals. During canal closure period the bed level measurements were done more intensively at relatively short distances in order to get an accurate assessment of deposition and scouring in the canal.

In this chapter water and sediment inflow rates to the irrigation system will be discussed. Water distribution in secondary offtakes and their delivery performance ratios also will be discussed. In sediment transport, the spatio-temporal sediment distribution, the sediment drawing efficiency of the secondary offtakes, the suspended sediment load and bed material particle size distribution and then the morphological changes in the canal cross-sections will be discussed.

6.1.   Methodology of field data collection

Data collection for the assessment of sediment transport in irrigation canals consists of mainly of two parts, the field measurements and the laboratory analysis. In field measurements the hydraulic measurements, suspended load and bed load sampling and morphologic measurements were performed. Whereas in laboratory analysis the sediment concentrations and particle size distribution etc, were determined from the samples collected during field measurements. Various kinds of data sets, their collection techniques and laboratory analysis are discussed in the subsequent sections.

There are four different kinds of data sets required for the assessment of sediment transport in irrigation canals.

Hydraulic data

- design discharges;
- canals' design longitudinal and cross-sectional data;
- structures' dimensions;
- actual discharges and velocities;
- hydraulic roughness;
- structures' calibration.

Sediment data

- sediment loads (spatio-temporal variation);
- sediment characteristics;
- suspended and wash load;
- bed material;
- boil samples.

Morphologic data

- variation in bed levels and cross sections.

Canal operation data

- actual and planned hydrograph at canal headworks during the study period;
- water distribution to secondary offtakes.

6.1.1. Measurement methods

Sampling points

A total number of seven sampling points, making seven reaches, were selected in the Machai Branch Canal and Maira Branch Canal for measuring water and sediment discharge along the canal. The locations of the sampling points were at head, middle and tail of the both canals and, upstream and downstream of the confluence and downstream of the head regulators of secondary offtakes. The water and sediment discharge measurement points are shown in Figure 6.1.

Hydraulics

The canals under study were divided into different reaches and then measurements regarding the water and sediment inflow and outflow were conducted. Depending upon the situation, the measurements were made directly by using discharge measuring equipments like current meters flumes and indirectly by water stage recording.

Water discharge in the main canals was measured directly by using a current meter. Depending upon the depth of water in the canal, the discharge was measured either by wading method or by boating. Cup type (Price AA) current meters were used for discharge measurements using International Sedimentation Research Institute (ISRIP) procedures (Mahmood et al., 1978). The cross-section was divided into 15-30 verticals, depending upon the water surface width. The widths and depths were measured along the tagline. At each vertical, two velocity measurements were made at points of 0.2 and 0.8 depth (for water depth greater than 1.5 m) and at the points of 0.6 depth only (for water depth less than 1.5 m). At every point the velocity was measured twice for two minutes, one minute per measurement and the average of two measurements was used for velocity determination. Figure 6.2 illustrates the current metering procedure.

This method of discharge measurement is called the mean section method. The cross-section is divided into subsections, each bounded by two adjacent verticals. If v_1 is the mean velocity at one vertical and v_2 is the mean velocity at the adjacent vertical, and d_1 and d_2 are the total depths measured at verticals 1 and 2, respectively, and b is the horizontal distance between verticals. The discharge in the vertical is obtained by multiplying the area of the subsection with the average velocity of the subsection. The area is obtained by multiplying the average depth to the width of the vertical. The discharge q_i of the subsection is given as:

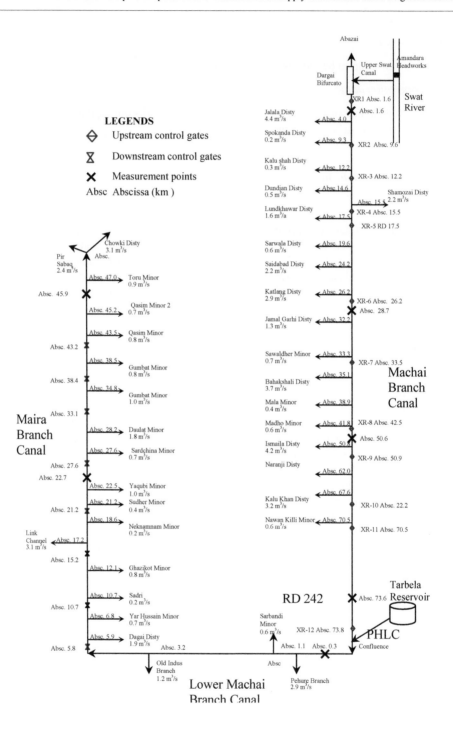

Figure 6.1. Water and sediment discharge measurement plan

$$q_i = \frac{U_{i-1} + U_i}{2} \frac{d_{i-1} + d_i}{2} b_i$$ (6.1)

This procedure is repeated in every sub section and total discharge passing through the cross-section is the summation of all subsections' discharges.

$$Q = \sum_{i=1}^{n} q_i$$ (6.2)

Where

U_i, U_{i-1}	= mean flow velocity (m/s)
b	= width of the subsection, m
d_i, d_{i-1}	= water depth at two verticals of the subsections i, m
q_i	= discharge in the subsection, m³/s
Q	= total discharge, m³/s
i	= number of subsection

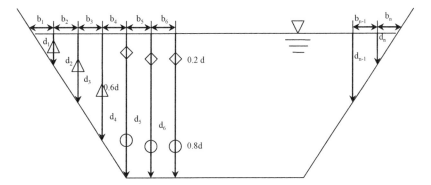

Figure 6.2. Observation points for discharge measurement in a canal cross-section

Accuracy of discharge measurement. Accuracy of discharge measurement depends on the number of verticals at which observations of depth and velocity are obtained. In general, the interval between any two verticals should not be greater than 1/20 of the total width, and the discharge between any two verticals should not be more than 10% of the total discharge. However, if the canal bed is very uniform 10 verticals may be adequate (Mahmood et al., 1978).

Discharges of the offtakes were measured with already installed crump weirs at the head of the offtakes. The discharge was estimated simply by reading the depth of water above the crump weir's crest. The depth of water was then converted into the discharge by using following formula:

$$Q = C_d B H^{3/2}$$ (6.3)

Where

Q = water discharge (m³/s)
C_d = discharge coefficient (m^{1/2}/s), 1.98
B = width of the weir (m)
H = height of water above weir crest (m)

Sediment sampling

Suspended load sampling. The depth integrated sediment sampling was conducted in the same canal cross-section selected for water discharge measurements. For suspended sediment sampling, the cross-section was divided into 10 verticals and 3 bottles from each vertical were collected, making 30 bottles in total. The verticals can be the same verticals as used for discharge measurement or different depending upon the shape of the canal bed, variation in velocity across the cross section and the amount of discharge in the canal as described in the paragraph relating to accuracy of discharge measurement. The water depths in almost all of the sample cross-sections were more than 1.5 m. Therefore the D-74 suspended sediment sampler was used with the Equi-Transit Rate (ETR) Method. For taking a representative sample the transit rates for all sampling locations were calculated before sampling.

     Equi Transit Rate (ETR) Method. A cross-section suspended sediment sample obtained by the ETR method requires a sample volume proportional to the amount of flow at each of several equal spaced verticals across the canal and an equal transit rate (ETR), both up and down in small verticals yields a gross sample proportional to the total canal flow. It is necessary to keep the same size nozzle in the sampler for a given measurement. This method is most often used in shallow or sand bed streams where the distribution of water discharge in the cross-section is not stable. The number of verticals required for an ETR sediment discharge measurement depends on the canal flow and sediment characteristics at the time of sampling.

     ETR Method means that (i) the sampling verticals are equally spaced; and (ii) all samples are taken at equal transit rate, as explained under:

(i). *equally spaced verticals.* The width of the segment which was sampled was determined by dividing the canal width by the number of 10 verticals as shown in Figure 6.3.B. Water surface width was determined from the tagline stretched across the canal.

(ii). *equal transit rate (ETR).* The maximum transit rate of the sampler (i.e. the speed at which the sampler is raised or lowered) must not exceed 0.4 of the mean flow velocity. Mean flow velocity was seen from the discharge measurement notes taken already during flow measurements. This transit rate was adequate to fill a bottle in a round trip and the speed of lowering and raising the sampler was kept same. The velocity of lowering and raising the bottle (the transit rate) was so arranged that sufficient sample was collected in each bottle and the bottle was not overfilled or less filled as shown in Figure 6.3.B. Three sample bottles were collected from each vertical. Thus it made a total of 30 bottles of suspended sediment samples for the 10 locations at one measurement site. In case of less filling or overfilling of the bottle the sample is discarded and the bottle is refilled to get an representative sample.

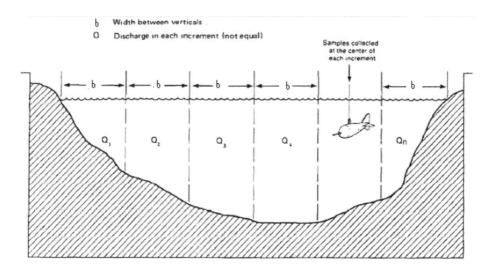

Figure 6.3.A. Equal width increment sampling technique

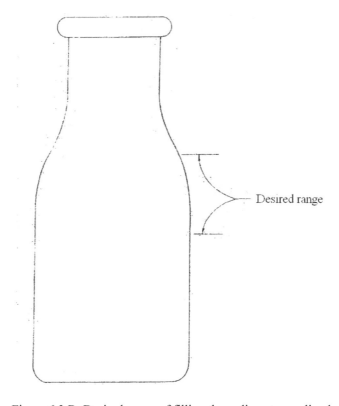

Figure 6.3.B. Desired range of filling the sediment sampling bottle

Bed material sampling

Bed material sampling was also done during detailed sediment sampling. Bed material samples were collected at three locations along the tagline at $T_w/6$, $T_w/2$ and $5T_w/6$, where T_w is the water surface width. Bed material samples were taken by using BM-54 bed material sampler. A total of four composite samples were collected. The samples were collected by following the USGS standard procedure (Guy, 1969).

Boil sampling

Boil samples were collected by using preferably DH-48 at the immediate downstream of canal structures, head regulators, cross-regulators and drop structures. The cross-section was divided into 2-5 sub-sections, depending upon the width of the cross-section. Then the samples were taken from various depths in every sub-section and finally a composite sample was prepared. The boil sampling was done at about twenty locations in the canal network, starting from Machai Branch Canal headworks to the tail of the Maira Branch Canal. Figure 6.4 shows a boil downstream of the head regulator of Machai Branch Canal.

Figure 6.4: Boil sampling location immediately downstream of a head regulator

Topographic observations

Water surface slope was determined by measuring actual water surface elevations in the study reach. The study reach was a 1,000-1,500 m long reach, at the water and sediment discharge measurement points. The water surface elevations and canal cross-sections were measured at 5 location in the study reach, using a dumpy level. The levels were then obtained from nearby benchmarks established for this purpose. Besides this, the canal cross-sections were measured at various other points in order to know the bed level variations along the canal and to assess the changes in initial and final bed levels. The initial bed levels were the bed levels at the time of starting the study. In this case, as the canal was cleaned in every January, so the design bed levels were considered as the initial bed levels were compared to the design bed levels. In January 2007 and 2008, the intensive cross-sectional and longitudinal survey were performed in order to assess the volume of sediment deposition/scouring and to determine any change in the canal bed slope and cross-section.

Laboratory analysis

For determining the particle size distribution and sediment concentration the sieve analysis, Visual Accumulation Tube (VAT) analysis and Pipette Method were used, depending upon the sediment sizes. The sediment analysis procedures of Guy (1969) were followed.

6.1.2. Fieldwork arrangement

Field data collection

Field data was collected for a total period of about two years, from October 2006 to September 2008, on the operation, sediment inflow, hydraulics and sediment transport in the canals under study.

Canal operation data

Canal operation data was collected from the Department of Irrigation (DoI), Swabi Division and Malakand Division. DoI has daily records of flow data. The daily flow data for the main canals (Machai Branch Canal, Maira Branch Canal and PHLC) and for the offtaking canals was collected for four cropping seasons *Rabi* 2006-2007, *Kharif* 2007, *Rabi* 2007-2008 and then *Kharif* 2008 again. The *Rabi* and the *Kharif* are two cropping seasons in a year in Pakistan. The *Rabi* season starts in mid October and ends in mid April and the *Kharif* season starts in mid April and ends in mid October.

Sediment inflow measurements

Sediment inflow measurements started in March 2007 at the headworks of Machai Branch Canal. During low sediment concentration periods (February to April and October to December) sediment samples were collected on fortnightly basis. During the peak sediment concentration period (May to September) weekly or biweekly sediment sampling was performed. Sediment sampling was also done at RD 242, Figure 6.4, in order to assess the sediment entry into the lower Machai Branch Canal and Maira Brach Canal on the same pattern.

Hydraulics and sediment transport measurements

In order to determine the temporal and spatial variability of flow and sediment content in the canal, the hydraulic and sediment transport measurements were conducted all along the canal at different time periods. A total number of three mass balance studies were conducted during the peak and low sediment concentration periods in July and December 2007 and in July 2008. In these measurements all of the water and sediment inflows and outflows were measured along with morphological changes.

6.2.   Irrigation water supply to USC-PHLC system

Figure 6.5 presents the average monthly discharges at the head of Machai Branch Canal, PHLC and at RD 242. The average flow supplied to Machai, PHLC and RD 242 during these four cropping seasons was 28.8 m^3/s, 12.4 m^3/s and 14.8 m^3/s respectively. The average flow supplies were much less than the design flow supplies. The design discharge of Machai Branch Canal, PHLC and at RD 242 is 66.7 m^3/s, 28.3 m^3/s and 31.9 m^3/s respectively.

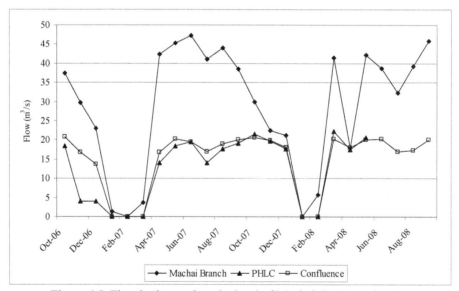

Figure 6.5. Flow hydrograph at the head of Machai, PHLC and at RD 242

Figure 6.6 presents seasonal water supplies to the irrigation canals under study, which shows the seasonal variations in the canal supplies. It is evident from Figure 6.6 that the supplies in the *Kharifs* are much higher than in the *Rabis*, due to high water demands and high water availability in the system.

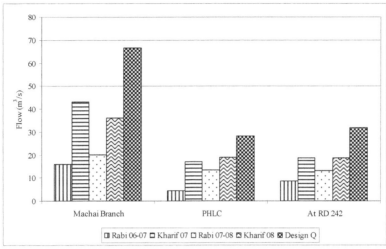

Figure 6.6. Seasonal water supplies to study irrigation canals

6.3.   Water distribution to secondary offtakes

Figure 6.7 presents average discharge supplies to the Machai Branch Canal offtakes up to RD 242 during the study period, where Figure 6.7 presents the average water supplied to the offtakes downstream of RD 242.

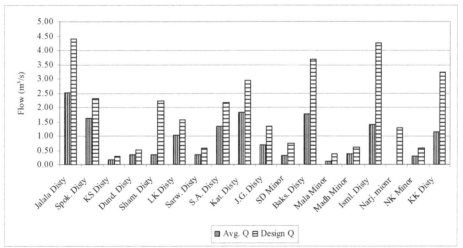

Figure 6.7. Average discharges of Machai Branch Canal offtakes from October 2006 to September 2008

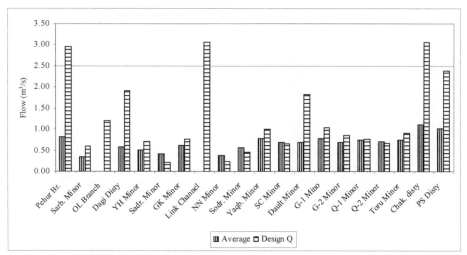

Figure 6.8. Average discharges of offtakes downstream of RD 242 from October 2006 to September 2008

Figure 6.7 and 6.8 show that almost every offtake was supplied with less discharge than the design. The reason can be less water availability in the Machai Branch Canal. The parent canal Upper Swat Canal (USC) of Machai Branch Canal receives water directly from the Swat River, without any storage. Therefore water availability in the canal is highly dependent upon the water availability in the river, which in winter is quite low and cannot feed the canal at its design discharge. Whereas in case of the irrigation system downstream of RD 242, the demand is generally lower than the design discharges of the offtakes. So the water users, usually, dispose water to nearby drains during low water demand periods and also close the tertiary outlets, which cause water losses on one hand and damage to canals due to overtopping on the other. Therefore, the Department of Irrigation (DoI) usually reduces the flows in winter periods.

Figures 6.9 and 6.10 present the Delivery Performance Ratio (DPR) of Machai Branch Canal system and the system downstream of RD 242 respectively. Similarly, the DPR values seem somehow low, mainly in Machai Branch Canal system and partly in RD 242 system. The reasons for lower DPR have already been discussed in the previous paragraph.

In Maira Branch Canal mainly smaller offtakes have higher DPR values as compared to larger offtakes, particularly in the system downstream of RD 242. The Sadri Minor and the Neknam Minor have design discharges 0.19 m³/s and 0.23 m³/s respectively and their DPR values were 2.0 and 1.6 respectively. It shows that the gate operators of the Department of Irrigation release more water than the design to these smaller offtakes on farmers demand. Generally a less risk of overtopping is associated with the smaller offtakes or there is not a big loss associated with the breach of these smaller canals. Hence the gate operators usually do not care about these factors while releasing high amounts of discharges in smaller offtakes.

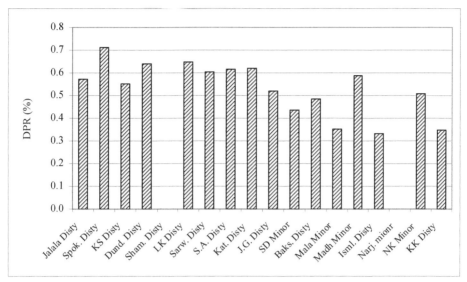

Figure 6.9. DPR of Machai Branch Canal offtakes

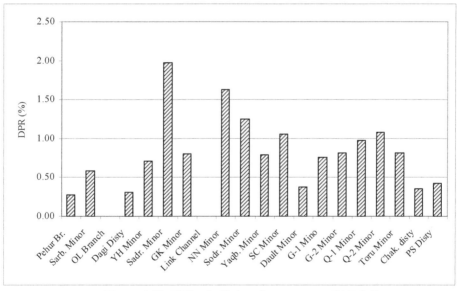

Figure 6.10. DPR of offtakes downstream of RD 242

6.4.   Sediment inflow to the system

Machai Branch Canal is an offtake of Upper Swat Canal (USC), whereas USC gets water from Swat River as mentioned earlier. USC has a vortex type silt ejector at its head. But this silt ejector is not functioning due to resistance from the riparian community living around the stream which carries water from the silt ejector. After travelling a distance of 6.1 km USC ends into Machai Branch Canal and Abazai Branch Canal at Dargai Bifurcator. Due to non-functionality of the silt ejector and direct diversion (without any storage) from the river, USC carries a handsome amount of sediment to the Machai Branch Canal and Abazai Branch Canal. Figure 6.11.A presents monthly average sediment concentrations at the head of the Machai Branch Canal during the study period. Figure 6.11.A depicts that the sediment entering into the system has the typical trend of sediment availability in the Indus Basin System of Rivers. It is higher in summer (monsoon) months and much lower in winter months. The fractions of sand, silt and clay in the total sediment concentration at Machai Branch Canal head are given in Table 6.1.A.

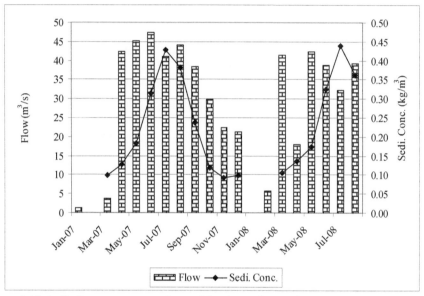

Figure 6.11.A. Sediment concentration and load entering to the system at Machai Branch Canal head

A same trend can be seen in Figure 6.11.B, which presents the monthly average sediment concentration and monthly sediment loads available at RD 242. The sediment load available at RD 242 was finer than the sediment load at the head of Machai Branch Canal. It shows that the coarse particles settle in upstream reaches and finer particles travel along the flow. Though, there are many other factors influencing the sediment transport in the canal, but generally, the flow control affects the sediment transport the most. To attain the required water levels in the canal ponding effects are generally created, which lead to sedimentation in the canals due to drop in flow velocities. Table 6.1.A and 6.1.B present the sediment characteristics entering into the Machai Branch Canal and at RD 242 respectively. Silt is the dominant factor, ranging from 60 to 70% in the sediment characteristics, sand is the other major portion, whereas clay concerns a small portion ranging between 10 to 20% of the total load.

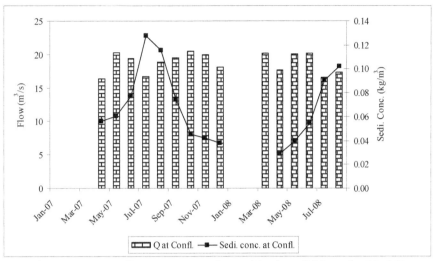

Figure 6.11.B. Sediment concentration and load available at the RD 242

Table 6.1.A. Flow, sediment concentration, sediment classes and sediment load at Machai Branch head

Month	Discharge	Sediment concentration	Sand >0.062 mm	Silt < 0.062 mm	Load
	(m^3/s)	(kg/m^3)	(kg/m^3)	(kg/m^3)	(Tonnes/month)
Jan-07	1.3				
Feb-07	0.0				
Mar-07	3.6	0.099	0.033	0.064	929
Apr-07	42.3	0.127	0.042	0.085	13,900
May-07	45.2	0.183	0.060	0.123	21,500
Jun-07	47.3	0.314	0.104	0.210	38,500
Jul-07	41.1	0.428	0.141	0.287	45,600
Aug-07	44.0	0.381	0.126	0.255	43,500
Sep-07	38.5	0.237	0.078	0.159	23,600
Oct-07	29.8	0.119	0.039	0.080	9,200
Nov-07	22.4	0.093	0.031	0.063	5,400
Dec-07	21.3	0.098	0.032	0.065	5,400
Jan-08	0.00				
Feb-08	5.7				
Mar-08	41.4	0.105	0.035	0.070	11,300
Apr-08	17.9	0.135	0.045	0.090	6,300
May-08	42.2	0.173	0.057	0.114	18,900
Jun-08	38.7	0.324	0.107	0.217	32,500
Jul-08	32.2	0.438	0.145	0.293	36,600
Aug-08	39.3	0.361	0.119	0.242	36,700

Table 6.1.B. Flow, sediment concentration, sediment classes and sediment load at RD 242 and at confluence

Month	Q at RD 242 (m³/s)	Q at conflu-ence (m³/s)	Concentra-tion at RD 242 (kg/m³)	Concent-ration at confluence (kg/m³)	Sand >0.062 mm (kg/m³)	Silt <0.062 mm (kg/m³)	Sediment load (Tonnes/month)
Jan-07							
Feb-07							
Mar-07							
Apr-07	6.5	16.4	0.197	0.056	0.004	0.196	2,400
May-07	6.5	20.3	0.249	0.060	0.007	0.246	3,200
Jun-07	7.7	19.4	0.271	0.077	0.009	0.267	3,900
Jul-07	7.9	16.7	0.396	0.128	0.012	0.312	5,500
Aug-07	8.0	18.9	0.389	0.116	0.015	0.384	5,700
Sep-07	8.2	19.6	0.253	0.075	0.011	0.249	3,800
Oct-07	6.5	20.5	0.189	0.045	0.005	0.188	2,400
Nov-07	6.5	19.9	0.172	0.042	0.004	0.171	2,200
Dec-07	6.2	18.1	0.149	0.038	0.003	0.148	1,800
Jan-08							
Feb-08							
Mar-08	6.3	20.2					
Apr-08	6.2	17.7	0.110	0.029	0.005	0.107	1,300
May-08	7.9	20.1	0.140	0.039	0.006	0.136	2,100
Jun-08	8.1	20.2	0.190	0.054	0.010	0.183	2,900
Jul-08	8.2	16.6	0.275	0.091	0.011	0.273	3,900
Aug-08	8.0	17.4	0.325	0.102	0.014	0.321	4,600

6.5.   Results from mass balance studies

The results of the mass balance studies are presented in the following.

6.5.1. Mass balance in July 2007

Suspended load

Table 6.2 presents the suspended sediment distribution along the canal. As mentioned earlier, the whole length of the canal was divided into seven reaches and all of the water and sediment inflow and outflow were measured in these reaches. It can be seen from Table 6.2 that silt is the dominant class in the suspended load. The discharge at the head of the Machai Branch Canal was 49 m³/s, which was 70% of the design discharge of the canal. The suspended particle size distribution (PSD) along the canals is given in Figure 6.12. It was observed that the median particle size decreases as it goes towards the tail. At head, the d_{50} is coarser and as it goes towards tail, the d_{50} decreases, where d_{50} is the median diameter. However, the d_{50} found at tail section of Maira Branch Canal was somewhat higher than the d_{50} found in the head and the middle. This pattern depicts that of the sediment entering into the canals, the coarser particles settle in the head reaches, whereas the fine particles travel along the canal. Then the scouring of the deposited material takes place in the middle reaches, due to which the coarser sediment was also found in the tail reaches.

Table 6.2. Suspended load distribution along the canal during July 2007

Date	Reach	Canal	Abscissa	> 0.062 mm	0.062 – 0.004 mm	<0.004 mm	Total concentr- ation
			(m)	(kg/m³)	(kg/m³)	(kg/m³)	(kg/m³)
16.07.07	R 1	Machai	1,585	0.012	0.054	0.006	0.072
16.07.07	R 2	Machai	28,708	0.009	0.084	0.013	0.106
17.07.07	R 3	Machai	50,637	0.003	0.202	0.017	0.223
17.07.07	R 4	Machai	73,453	0.005	0.175	0.014	0.194
18.07.07	R 5	Lower Machai	600*	0.003	0.034	0.013	0.050
18.07.07	R 6	Maira	22,713*	0.000	0.071	0.010	0.080
19.07.07	R 7	Maira	45,880*	0.010	0.061	0.020	0.092

*Abscissa from RD 242

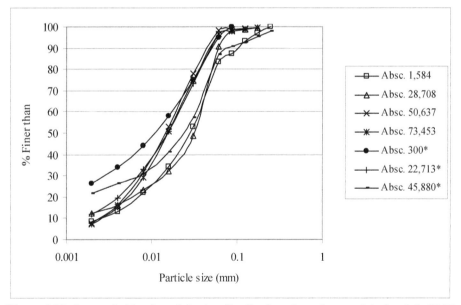

Figure 6.12. Suspended load particle size distribution along the canal during July 2007

Bed material size distribution

Similar trend was found in the bed material size distribution, as the coarser d_{50} at head and tail reaches and finer d_{50} at the middle reaches was found. The d_{50} at head reach of Machai Branch Canal was 0.296 mm, whereas in the middle reaches, the d_{50} was 0.131

and 0.102 mm. The d_{50} at tail of Machai Branch was also higher, that was 0.198 mm. Whereas in Maira Branch Canal the d_{50} at head and middle reaches was larger than the d_{50} at tail reach. At head and middle reaches the d_{50} was 0.212 mm and 0.130 mm, whereas at tail reaches the d_{50} was 0.134 mm. Table 6.3 presents the bed material size distribution along the canal in July 2007.

Table 6.3. Bed material size distribution along the canal during July 2007

	Date	Canal	Abscissa	D_{16}	D_{35}	D_{50}	D_{65}	D_{84}	D_{90}	Sigma
			(m)	(mm)	(mm)	(mm)	(mm)	(mm)	(mm)	
1	16.07.07	Machai	1,585	0.209	0.26	0.296	0.332	0.398	0.430	1.38
2	16.07.07	Machai	28,708	0.110	0.124	0.131	0.144	0.188	0.233	1.31
3	17.07.07	Machai	50,637	0.077	0.089	0.102	0.119	0.175	0.250	1.52
4	17.07.07	Machai	73,453	0.117	0.16	0.198	0.242	0.258	0.312	1.37
5	18.07.07	L. Machai	600[*]	0.167	0.194	0.212	0.232	0.268	0.290	1.26
6	18.07.07	Maira	22,713[*]	0.065	0.101	0.130	0.173	0.21	0.250	1.59
7	19.07.07	Maira	45,880[*]	0.055	0.099	0.134	0.176	0.191	0.203	1.00

[*] Abscissa from RD 242

Total load

Table 6.4 presents the overall water and sediment mass balance during July 2007 exercise. In total 681 tonnes sediment entered into the system and an amount of 370 tonnes was withdrawn by the secondary offtakes and direct outlets, whereas, the balance of 258 tonnes was deposited in the canal, which is about 37% of the incoming load. There are certain canal reaches, which are showing a sediment deposition trend and some an erosion trend. In Machai Branch Canal the middle reach, from RD 28,650 to 50,716, showed an erosion trend, where the rest of the reaches showed a deposition trend as given in Table 6.4.

Table 6.4. Water and sediment balance during July 2007 measurement campaign

Date	Reach	Canal	Abscissa (m)		Q-in	C-in	Q-out	C-out	Dep/Ero
			From	To	(m^3/s)	(kg/m^3)	(m^3/s)	(kg/m^3)	(T/day)
16.07.07	R1	Machai	1,585	28,650	44.8	0.180	24.3	0.17	170
16.07.07	R2	Machai	28,650	50,716	24.3	0.170	15.0	0.47	-341
17.07.07	R3	Machai	50,716	73,453	15.0	0.470	6.6	0.20	423
17.07.07	R4	Machai	73,453	600[*]	6.6	0.200	24.9	0.06	-14
18.07.07	R5	Lower Machai	600[*]	22,713[*]	24.9	0.060	11.9	0.10	-19
18.07.07	R6	Maira	22,713[*]	45,880[*]	11.9	0.100	5.2	0.12	20
19.07.07	R7	Maira	45,880[*]	48,020[*]	5.2	0.120	5.0		19
Net Deposition									258

*Abscissa from RD 242

6.5.2. Mass balance in December 2007

Suspended load

As July is the monsoon month, generally higher sediment concentrations come into the river flows. Contrary to July, December is a winter month and river flows are relatively clear in winter months. Therefore, generally, canals do not run at their design discharges because of low flow in the rivers and low crop water requirements of the command area. To see the effects of these phenomena on the sediment transport, the other measurement campaign was undertaken in December 2007. The discharge at the head of Machai Branch Canal during this campaign was 25.5 m^3/s, which was 40% of the design discharge of Machai Branch Canal.

Table 6.5 presents the results of suspended load distribution along the Machai Branch and Maira Branch canals during December 2007. The suspended load concentrations were lower than in July 2007. As in July 2007 the suspended sediment concentration entering into the Machai Branch Canal was 0.072 kg/m^3, whereas in this month the sediment entering into the canal was 0.056 kg/m^3. The dominant class in both July and December was the silt. As in July 2007 the silt was 75% whereas in December 2007 the silt was 82% of the total suspended load. The second major class was the sand and the clay was found in less quantity, about 8% and 5% in July 2007 and December 2007, respectively.

Table 6.5. Suspended load distribution during December 2007

Date	Reach	Canal	Abscissa	> 0.062 mm	0.062 – 0.004 mm	< 0.004 mm	Total
			(m)	(kg/m^3)	(kg/m^3)	(kg/m^3)	(kg/m^3)
12.12.07	R 1	Machai	1,524	0.007	0.046	0.003	0.056
13.12.07	R 2	Machai	28,680	0.011	0.083	0.004	0.098
14.12.07	R 3	Machai	50,639	0.001	0.076	0.009	0.085
15.12.07	R 4	Machai	73,392	0.003	0.088	0.007	0.097
16.12.07	R 5	Lower Machai	539[*]	0.010	0.108	0.007	0.126
17.12.07	R 6	Maira	22,652[*]	0.000	0.068	0.005	0.073
18.12.07	R 7	Maira	45,817[*]	0.019	0.078	0.017	0.114

*Abscissa from RD 242

Figure 6.13 presents the particle size distribution along the canal during December 2007 sediment measurement campaign. The median diameter found at head and middle reaches of Machai Branch Canal was 0.032 mm, whereas in tail reaches, the median diameter was 0.016 mm. The median diameter at head and middle reaches of Maira Branch Canal was 0.016 mm, whereas at tail reaches the median diameter was 0.031 mm. Particle sizes found in the December measurements were finer than in the July measurements. July is the monsoon season and majority of sediments come from the river basins with runoff. During erosion process form the river basins a large variety of detached sediments enter into the rivers, which is then diverted to the irrigation canals. These particles mainly consist of sand, silt and clay also. Therefore in July the particle size was comparatively larger than the particle sizes found in December measurements.

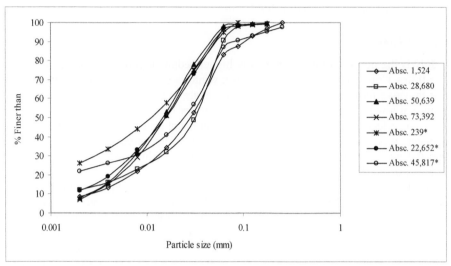

Figure 6.13. Suspended particle size distribution along the canal during December 2007

Bed material size distribution along the canal

Table 6.6 presents the bed material size distribution along the canal. The d_{50} at head reach was found to be 0.239 mm, whereas in middle reaches it was 0.169 and 0.192 mm. At tail reach of Machai Branch Canal the d_{50} was 0.167. In Maira Branch Canal the d_{50} at head was 0.084, at middle 0.110 and at tail 0.078 mm. Though the trend of bed material size distribution is almost similar to the July 2007 in Machai Branch Canal, but the size of d_{50} was rather smaller in December 2007 than in July 2007. In Maira Branch Canal the bed material d_{50} at middle and tail reaches was smaller than the head, which shows that the coarser particles settled from head to tail.

Table 6.6. Bed material particle size distribution during December 2007

	Date	Canal	Abscissa	D_{16}	D_{35}	D_{50}	D_{65}	D_{84}	D_{90}	Sigma
			(m)	(mm)	(mm)	(mm)	(mm)	(mm)	(mm)	
1	12.12.07	Machai	1,524	0.185	0.218	0.239	0.26	0.298	0.314	1.27
2	13.12.07	Machai	28,680	0.11	0.14	0.169	0.201	0.268	0.320	1.56
3	14.12.07	Machai	50,639	0.101	0.148	0.192	0.250	0.260	0.305	1.89
4	15.12.07	Machai	73,392	0.081	0.128	0.167	0.208	0.213	0.261	1.44
5	16.12.07	Lower Machai	539[*]	0.07	0.076	0.084	0.092	0.119	0.140	1.31
6	17.12.07	Maira	22,652[*]	0.071	0.092	0.110	0.130	0.158	0.178	1.28
7	18.12.07	Maira	45,817[*]	0.046	0.063	0.078	0.098	0.139	0.158	1.28

[*] Abscissa from RD 242

Total load in December 2007

Table 6.7 presents the water and sediment mass balance during December 2007 campaign. Overall a sediment deposition trend was observed. Though, the sediment concentration was less than in July 2007, but due to low flow, the hydrodynamic forces required to transport the sediment were also low. Head and middle reaches of Machai Branch Canal exhibited the deposition trend, whereas the two tail reaches showed scouring behaviour. Whereas in Maira Branch Canal, the head reach and the tail reach showed a depositing trend, whereas, scouring was observed in the middle reach. Overall, 153 tonnes of sediment was deposited in the canals, which is about 60% of the incoming load.

Table 6.7. Water and sediment mass balance during December 2007 campaign

Date	Reach	Canal	Reach		Q-in	C-in	Q-out	C- Out	Deposit-ion.
			From	To	(m^3/s)	(kg/m^3)	(m^3/s)	(kg/m^3)	(T/day)
12.12.07	R1	Machai	1,585	28,650	25.5	0.106	13.20	0.109	81
13.12.07	R2	Machai	28,650	50,716	13.2	0.109	7.90	0.086	56
14.12.07	R3	Machai	50,716	73,453	7.9	0.086	5.49	0.117	-3
15.12.07	R4	Machai	73,453	600*	5.5	0.117	5.49	0.117	-117
16.12.07	R5	Lower Machai	600*	22,713*	15.2	0.131	8.52	0.073	103
17.12.07	R6	Maira	22,713*	45,880*	8.5	0.073	4.16	0.137	-6
18.12.07	R7	Maira	45,880	48,020*	4.2	0.137			38
Net Deposition									153

* Abscissa from RD 242

6.5.3. Mass balance in July 2008

In July 2008 the sediment transport measurements were done at some low canal discharges but a relatively higher sediment concentration. The campaign was conducted from 22 July 2008 to 26 July 2008.

Suspended load distribution

Results of the suspended sediment distribution are given in Table 6.8. The suspended sediment concentration entering into the irrigation canal at the head of the Machai Branch Canal was 0.114 kg/m^3, with silt as a dominant class, consisting of 72% of the total suspended load, whereas the sand was found 18%, which is almost similar as in July 2007. The sediment distribution was almost like in July 2007, especially in Maira Branch Canal, where concentration increases towards the tail.

In July 2008 the water discharge in the canal was less than the discharge in July 2007. In July 2007 the water discharge was 44.7% m^3/s, which was almost 70% of the flow, whereas in July 2008 the discharge in the canal was 28.7 m^3/s, which was 43% of the design discharge. On the other hand the suspended load concentration in the flow was higher than the concentration in the July 2007. Therefore more deposition was observed during this campaign, particularly at the head reaches.

Table 6.8. Suspended sediment load distribution in irrigation canals

Date	Reach	Canal	Abscissa (m)	>0.062 mm (kg/m³)	0.062–0.004 mm (kg/m³)	< 0.062 mm (kg/m³)	Total (kg/m³)
22.07.08	R 1	Machai	2,510	0.021	0.083	0.010	0.114
22.07.08	R 2	Machai	27,278	0.036	0.164	0.019	0.219
23.07.08	R 3	Machai	50,731	0.021	0.159	0.040	0.220
23.07.08	R 4	Machai	73,453	0.004	0.216	0.042	0.262
26.07.08	R 5	Lower Machai	600*	0.001	0.022	0.002	0.025
26.07.08	R 6	Maira	22,713*	0.004	0.060	0.014	0.078
26.07.08	R 7	Maira	45,880*	0.004	0.066	0.015	0.085

*Abscissa from RD 242

The suspended sediment particle size distribution during July 2008 is given in the Figure 6.14. The median diameter found at the head of Machai Branch Canal was coarser, but then gradually the median diameter got finer towards the tail of the canal. This time a somewhat different PSD was observed. The maximum particle size at the Machai Branch Canal head was found to be 0.025 mm but with a very small percentage. These larger sediments seem settled in the head reach, because the sediments got finer and finer towards the tail.

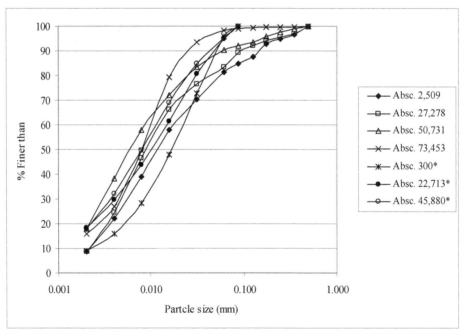

Figure 6.14. Particle size distribution along the canal during July 2008 measurements

Bed material size distribution

Table 6.9 presents bed material particle size distribution in July 2008. This time the larger particles were found in the middle and tail reaches of Machai Branch Canal, and tail reaches of Maira Branch Canal. The head reaches of Machai Branch Canal and Maira Branch Canals have finer bed material particles as compared with tail reaches, which was 0.192 and 0.136 respectively for Machai and Maira Branch Canals.

Table 6.9. Bed material size distribution along the canal during July 2008

Date	Reach	Canal	Abscissa	D_{16}	D_{35}	D_{50}	D_{65}	D_{84}	D_{90}	Sigma
			(m)	(mm)	(mm)	(mm)	(mm)	(mm)	(mm)	
16.07.08	R 1	Machai	1,585	0.125	0.149	0.192	0.298	0.387	0.400	1.78
16.07.08	R 2	Machai	28,708	0.109	0.135	0.165	0.209	0.245	0.303	1.81
17.07.08	R 3	Machai	50,637	0.121	0.145	0.161	0.189	0.243	0.278	1.42
17.07.08	R 4	Machai	73,453	0.099	0.108	0.131	0.142	0.195	0.225	1.30
18.07.08	R 5	Lower Machai	600*	0.086	0.116	0.136	0.151	0.172	0.205	1.27
18.07.08	R 6	Maira	22,713*	0.076	0.089	0.112	0.124	0.133	0.190	1.45
19.07.08	R 7	Maira	45,880*	0.070	0.125	0.158	0.195	0.208	0.210	1.33

*Abscissa from RD 242

Total load

Table 6.10 presents the total load transport during July 2008 sediment measurement campaign. The total sediment entered into the Machai Brach Canal was 654 tonnes, with a total concentration of 0.264 kg/m³. As the concentration of the sediments was higher at relatively low flow, so as a result as higher sediment deposition took place. A total of 349 tonnes sediment was deposited, which is 53% of the incoming load. Deposition volume in July 2008 was higher than the volume deposited in July 2007 and December 2007 net deposition.

Table 6.10. Water and sediment mass balance during July 2008

Date	Canal	Abscissa (m)		Q-in	C-in	Q-out	C-out	Deposition
		From	To	(m³/s)	(kg/ m³)	(m³/s)	(kg/ m³)	(T/day)
22.07.08	Machai	1,585	28,650	28.7	0.264	15.7	0.303	139
22.07.08	Machai	28,650	50,716	15.7	0.303	12.2	0.271	92
23.07.08	Machai	50,716	73,453	12.2	0.271	8.1	0.486	-103
23.07.08	Machai	73,453	600*	8.1	0.486	8.1	0.163	34
24.07.08	Lower Machai	600*	22,713*	21.7	0.163	11.1	0.880	163
25.07.08	Maira	22,713*	45,880*	11.1	0.880	6.0	0.960	2
26.07.08	Maira	45,880*	48,020*	6.0	0.960			21
Net Deposition								349

*Abscissa from RD 242

It has been observed that the amount of discharge in the canal and sediment concentration have great effect on overall sediment deposition pattern in the canal. Generally at low flows more sediment deposition took place and at higher flows less sedimentation took place, if there is enough sediment concentration in the flow. At low flows with less sediment concentration, not much deposition was observed. Further downstream controlled irrigation canals are more prone to sediment deposition than the upstream controlled irrigation canals. For example, in Machai Branch Canal less deposition was observed even at low flows, whereas in Lower Machai Branch Canal and the Maira Branch Canal more sediment deposition was found at the same amount of flows. It is because in downstream controlled irrigation canals the ponding effect starts with the reduction in flow. As far as the amount of flow goes down, the water level in the canal rises up correspondingly.

In July 2007, December 2007 and July 2008 the water discharge in the Machai Branch Canal was 69%, 40% and 45% of the design discharge and sediment deposition of 35%, 7% and 25% took place, whereas in Maira Branch Canal the water discharge was about 80%, 48% and 68% of the design discharge and sediment deposition of 15%, 80% and 69% took place respectively. Generally at higher flows the sediment transport capacity of the irrigation canals is generally higher, which is true in case of Lower Machai and Maira Branch Canals but shows opposite behaviour for Machai Branch Canal. It means the sediment transport in the canal not only depends upon the amount of discharge but also depends upon the amount of sediments in the flow and operation of the flow regulating structures in the canals.

6.6.   Offtakes sediment withdrawal

Offtakes sediment withdrawal efficiency is the key parameter in controlling sedimentation in the canal prism. Special attention is paid while designing and installing the offtakes' head regulators at the canal so that they can draw their fair share of sediment. Generally it is desired that the offtakes must draw the same concentration of the sediments as in the parent canal. The offtake sediment withdrawal is determined by the sediment withdrawal efficiency of the offtakes, which is the ratio of the sediment concentration in the offtaking canal to the sediment concentration in the parent canal. If sediment concentration of the offtaking canal is equal to the parent canal then the offtakes' sediment withdrawal efficiency is said to be 100%.

Many factors affect the sediment withdrawal efficiency of the offtakes. The major factors are the suspended sediment distribution and the amount of the bed load. If suspended load is higher, then the sediment withdrawal efficiency will be higher. The other factors affecting the sediment withdrawal efficiency include the crest levels of the offtakes, water levels in the canal, mean sediment concentration, distance from the cross regulating structures and then the dimensions, size and angle of installation of the offtakes head regulators and downstream water levels. Some salient information which may have important influence on offtakes sediment withdrawal efficiencies is given in Table 6.11. In this table the design discharge, actual discharges, DPR values, height of sill levels with reference to canal bed level and normal water level are given.

Table 6.11. Salient information of offtakes of Machai Branch Canal

Abscissa (m)	Offtake	Design Q (m³/s)	Actual Q (m³/s)	DPR	Sill level – bed Level (m)	Water level – crest level (m)
3,993	Jalala Disty	4.4	4.4	1.01	1.03	1.58
9,280	Spokanda Disty	2.3	2.4	1.03	0.25	2.26
12,227	Kalu Shah Disty	0.2	0.3	1.16	0.76	1.69
14,657	Dundian Disty	0.5	0.7	1.33	0.31	2.14
15,483	Shamozai Disty	2.2	2.2	0.97	1.11	1.32
17,538	Lund Khawar Disty	1.6	1.6	1.02	0.99	1.38
19,556	Sarwala Disty	0.6	0.6	1.11	0.38	2.45
24,152	Saidabad Disty	2.2	2.3	1.05	0.62	1.89
26,187	Katlang Disty	2.9	3.1	1.05	0.82	1.67
32,227	Jamal Garhi Disty	1.3	1.4	1.03	0.30	2.97
33,289	Sawal Dher Minor	0.7	0.6	0.75	0.15	2.42
35,128	Bhakshali Disty	3.7	2.9	0.79	0.61	1.60
38,877	Mala Minor	0.4	0.3	0.93	0.61	1.48
41,760	Madho Minor	0.6	0.5	0.77	0.61	1.46
50,808	Ismaila Disty	4.3	3.0	0.71	0.61	1.61
62,033	Naranji Minor	1.3	0.7	0.55		
67,569	Kalu Khan Disty	3.2	2.5	0.76	0.61	1.42
70,518	Nawan Killi Minor	0.5	0.4	0.76	0.61	1.30
1,425	Pehure Branch	2.9	1.7	0.59	1.23	0.72
1,870	Sarbandi Minor	0.6	0.7	1.16	1.08	0.9
3,470	Old Indus Branch	1.2	0.7	0.58	1.22	0.72
5,910	Dagi Disty	1.9	1.7	0.87	0.59	1.08
6,800	Yar Hussain Minor	0.7	0.6	0.89	0.87	0.9
10,740	Sadri Minor	0.2	0.3	1.40	0.94	0.69
12,110	Ghazi Kot Minor	0.8	1.0	1.35	0.84	0.9
17,170	Link Channel	3.1	1.8	0.58	0.56	1.26
18,595	Nek Nam Minor	0.2	0.4	1.55	0.99	0.54
21,230	Sudher Minor	0.4	0.7	1.65	0.97	0.72
22,490	Yaqubi Minor	1.0	1.0	1.00	0.76	0.89
27,630	Sard China Minor	0.7	0.6	0.97	0.63	0.72
28,160	Daulat Minor	1.8	1.2	0.66	0.23	1.08
34,760	Gumbat-1 Minor	1.0	0.8	0.81	0.32	0.9
38,490	Gumbat – 2 Minor	0.8	0.7	0.84	0.34	0.84
43,530	Qasim-1 Minor	0.8	1.1	1.45	0.32	0.78
45,210	Qasim-2 Minor	0.7	0.8	1.21	0.38	0.72
46,980	Toru Minor	0.9	0.8	0.88	0.39	0.9
48,020	Pir Sabaq Disty	2.4	1.7	0.73	0.59	0.76
48,020	Chauki Disty	3.1	2.4	0.79	0.54	0.81

All these factors play an important role in the sediment withdrawal efficiency of offtakes. Anyhow it was not possible to determine the effect of all of these parameters during the field measurements in this study. It needs more detailed studies to develop relationships among different parameters and offtakes sediment withdrawal efficiency. Amongst these parameters the height of the crest level plays the most important role. As bed load usually moves in layers close to bed and offtakes having lower sill levels receive more bed load and consequently receive higher sediment discharge and attain higher sediment withdrawal efficiency. The flow conditions in the canal also influence sediment withdrawal efficiency. The offtakes which are close to the cross regulators in the upstream direction also have less sediment withdrawal efficiencies. Because if the cross regulators are not properly operated then offtakes close to the cross regulators have the ponding effect due to backwater created at the cross regulators. In this case the water depth becomes higher than the normal depth and flow velocities become lower. Resultantly the amount of sediments in suspension becomes less and a less amount of sediment is discharged into the offtaking canals.

Tables 6.12 presents the results of sediment drawing efficiency of secondary offtakes and direct outlets in the Machai Branch canal system and RD 242 System respectively. It has been observed that the offtakes have higher efficiencies in summer months than in winter months, or in other words, during high sediment concentration, the offtakes sediment withdrawal efficiencies are comparatively higher. Overall the offtakes have lower efficiencies, which might be due to a high amount of bed load. As almost 50% of the total load moves in the form of bed load. This portion of total load then becomes difficult to be discharged into the offtakes which have higher sill levels.

The Machai Branch Canal offtakes have higher sediment withdrawal efficiencies as compared to Maira Branch offtakes. It can be attributed to a number of factors like the relatively lower crest levels in Machai Branch Canal as compared to the Lower Machai Branch Canal and the Maira Branch Canal and high ratio of suspended load in Machai Branch Canal. Further the Machai Branch Canal is a supply based and upstream controlled irrigation canal where flow velocities are usually higher than the semi-demand based downstream controlled Lower Machai Branch Canal and Maira Branch Canal. The higher sediment withdrawal efficiencies of Machai Branch Canal can also be a reason of low sediment deposition in the Machai Branch Canal.

No significant difference was observed in sediment withdrawal efficiencies along the distance in Machai Branch Canal and Maira Branch Canal. There was a mixed trend of the withdrawal efficiencies of the offtakes at the head, middle and tail of these canals. However the offtakes having low design discharges have greater sediment withdrawal efficiencies. It was observed that the sediment withdrawal efficiencies were higher when the flow in the parent canal was higher and were less when the flow was less in both upstream controlled and downstream controlled irrigation canals. It may be because in case of higher flows the amount of sediments in suspension are generally higher and more sediment is diverted to the offtaking canals. In downstream controlled irrigation canals the offtakes closed to cross regulators at the downstream side have higher sediment withdrawal efficiencies as in case of Dagi Distributary, Sadri Minor and Qasim-1 Minor.

Table 6.12. Sediment drawing efficiency in of Machai Branch Canal offtakes

	Abscissa (m)	Offtake	July 2007	December 2007	July 2007	Average
			%	%	%	%
1	3,993	Jalala Disty	85	25	59	56
2	9,280	Spokanda Disty	55	38	61	51
3	12,227	Kalu Shah Disty	56	23	65	48
4	14,657	Dundian Disty	61	26	52	47
5	15,483	Shamozai Disty	77	32	59	56
6	17,538	Lund Khawar Disty	70	25	52	49
7	19,556	Sarwala Disty	59	23	52	44
8	24,152	Saidabad Disty	70	32	72	58
9	26,187	Katlang Disty	82	26	82	63
11	32,227	Jamal Garhi Disty	58	24	83	55
12	33,289	Sawal Dher Minor	92	26	76	64
13	35,128	Bhakshali Disty	64	26	66	52
14	38,877	Mala Minor	95	33	61	63
16	4,1760	Madho Minor	58	31	44	44
18	50,808	Ismaila Disty	57	42	73	57
19	62,033	Naranji Minor	58	35	38	44
20	67,569	Kalu Khan Disty	49	37	79	55
21	70,518	Nawan Killi Minor	65	44	62	57
1	1,425	Pehure Branch	75	21	37	45
2	1,870	Sarbandi Minor	78	27	36	47
3	3,470	Old Indus Branch	60	21	41	41
4	5,910	Dagi Disty	72	42	67	60
5	6,800	Yar Hussain Minor	67	17	36	40
6	10,740	Sadri Minor	72	24	48	48
7	12,110	Ghazi Kot Minor	57	17	36	37
8	17,170	Link Channel	72	20	50	47
9	18,595	Nek Nam Minor	87	24	54	55
10	21,230	Sudher Minor	107	31	55	64
11	22,490	Yaqubi Minor	85	24	42	50
13	27,630	Sard China Minor	62	48	129	80
14	28,160	Daulat Minor	45	38	45	43
15	34,760	Gumbat-1 Minor	61	38	106	68
16	38,490	Gumbat – 2 Minor	61	43	137	80
17	43,530	Qasim-1 Minor	70	51	125	82
18	45,210	Qasim-2 Minor	61	41	149	84
20	46,980	Toru Minor	61	18	35	38
21	48,020	Pir Sabaq Disty	66	25	64	52
22	48,020	Chauki Disty	66	25	64	52

6.7.   Morphological changes in the canals

To determine the morphological behaviour of the canals under study, a total number of four cross-sectional surveys were conducted. First cross-sectional survey was conducted in January 2007, second in July 2007, third in January 2008 and last one in July 2008. The measured bed levels in these surveys were compared with the design bed levels and with other surveys in order to see the morphological behaviour of the canal. The results of these surveys are given in detail in the following:

6.7.1. Cross-sectional survey in January 2007

Figure 6.15 gives the results of cross-sectional survey of January 2007. The sedimentation took place mainly in the head reach. There is about 300 metre distance between the cross regulator at RD 242 and the confluence, where PHLC falls into the Machai Branch Canal (is called the lower Machai Branch Canal). Due to incoming flow from PHLC, ponding takes place here and a large proportion of sediments drops into this area. Therefore, more than one metre bed level raise was observed here. Anyhow, the canal was desilted and the cross-sections were brought back at the design values in January 2007.

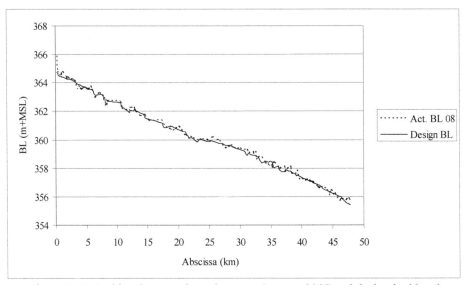

Figure 6.15. Bed level comparisons between January 2007 and design bed level

6.7.2. Cross-sectional survey in July 2007

Figure 6.16 presents the bed level variations from March 2007 to July 2007. Canal came into operation in late March and these variations can be attributed to four months. This survey shows a mixed trend of deposition and erosion. It is difficult to measure the cross-sections in flowing water thus only at limited and important points the cross-sections were measured. On overall 80 cross-sections were surveyed, 45 in Machai

Branch Canal and 35 in Maira Branch Canals. The cross-sections were measured particularly at the upstream and downstream of the cross regulators and in the middle of two cross regulators.

The variations in bed levels presented in the Figure 6.16 have been compared with the design bed levels. In both canals the desiltation exercise took place in January 2007 and bed levels were restored to the design bed levels. It can be observed that the variations in the canal bed were higher in the Machai Branch Canal than the Maira Branch Canal. It might be due to more flow control and low flows in the Maira Branch Canal. It can be observed that the variations decreased along the distance. The discharges at the head reaches of the Machai Branch Canal are also higher and decrease gradually towards the tail. The higher discharges exert more force at the canal bed and may lead to degradation phenomenon.

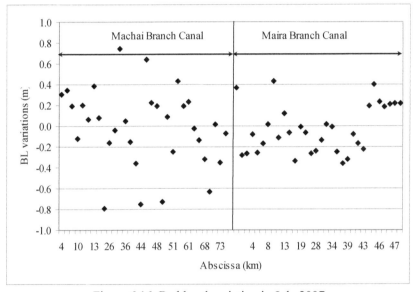

Figure 6.16. Bed level variation in July 2007

6.7.3. Cross-sectional survey in January 2008

Figure 6.17 presents cross-sectional survey in January 2008. On overall there is not a serious sedimentation problem in the Maira Branch Canal, except for the first 300 metres at the start of the canal. As, already mentioned, at this location, ponding effects remains dominant due to high inflow from PHLC at the confluence point, which is at an abscissa of 300 metres from the RD 242. Sediment coming from PHLC, drops here, in relatively large amounts. But this deposition does not generally obstruct the flow, as the section is wide enough to accommodate high amounts of discharge. In middle reaches, the scouring trend was observed, whereas, at the tail, again some deposition took place. It shows that after some deposition at the head reaches, the relatively clear water pick sediment in the middle reaches and then again at the tail, the discharge becomes low and hence hydrodynamic forces are not enough to carry the sediments. The phenomenon causes deposition at the tail sections.

During the study period the morphology of the canal remained changing, but overall there was not a severe sedimentation problem. As if compared with the sediment transport capacity of the canal, the incoming sediment was quite less than the transport capacity. But, still a sediment deposition trend was found, which might be due to flow control mechanism in the canal. During January 2008 canal cross-sectional survey a total volume of 19,300 m^3 sediment deposition was found. Whereas in July 2007 and July 2008 only the bed level and cross-sectional variations were determined.

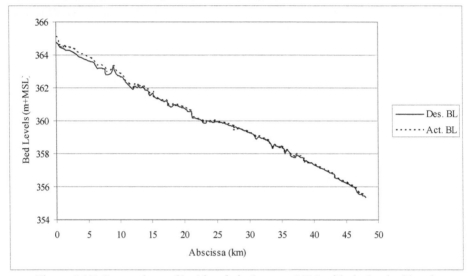

Figure 6.17. Comparison of bed levels in January 2008 with design bed levels

7. Canal operation and sediment transport

7.1. Background

After collecting field data on flow and sediment transport in the irrigation canals under study, sediment transport modelling was performed in order to assess further scenarios from sediment management perspective under various hydraulic and sediment inflow conditions. The main purpose of the modelling was to assess the effect of different operation schemes on sediment transport in the demand based automatically downstream controlled irrigation canals after calibrating and validating some existing sediment transport model.

A large number of sediment transport models are available for flow and sediment transport simulations in open channels, but only a few of them can be used for simulating sediment transport in irrigation canals as discussed in the chapter on literature review. In this study a one dimensional hydraulic and sediment transport model Simulation of Irrigation Canals (SIC) Model has been used. The detailed description of the model is given in the subsequent sections.

7.2. Simulation of Irrigation Canals (SIC) Model

Simulation of Irrigation Canals Model consists of separate hydraulic and sediment modules to deal with flow and sediment transport modelling in irrigation canals (Baume, 2005). It is a one dimensional (1-D) mathematical model, which allows the simulation of hydraulic behaviour of irrigation canals, in steady and unsteady state flow conditions. The sediment module of the SIC Model permits simulation of sediment transport in irrigation canals. It is capable of simulating the suspended sediment distribution in the cross-section, the sediment deposition and scouring, the bed formation, the sediment diversions to the offtaking canals and bed grain size distribution both in steady state and unsteady state flow conditions. SIC model comprises of the following four modules as:

- Topography Module;
- Steady Flow Module;
- Unsteady Flow Module;
- Sediment Module.

7.2.1. Topography module

In the topography module the canal geometrical data is entered after which the geometrical calculations are performed which are used in the steady and unsteady flow modules.

7.2.2. Steady flow module

The steady flow module computes the water surface profile in the canal under steady flow conditions. It allows to test the influence of modifications to the structures and canal maintenance. It also computes the offtake gate openings to satisfy given target discharges and the cross-regulator gate openings to obtain a given target water surface elevation upstream of the regulator.

Differential equation for water surface profile. The differential equation of a water surface profile can be written as:

$$\frac{dH}{dx} = -S_f + (k-1)\frac{qQ}{gA^2} \tag{7.1}$$

with:

$$S_f = \frac{n^2Q^2}{A^2R^{4/3}}$$

Where

g	= acceleration due to gravity (m/s^2)
n	= Manning's roughness coefficient (s/m$^{1/3}$)
R	= hydraulic radius (m)
A	= cross-sectional area (m)
H	= total head (m)
q	= lateral inflow/outflow (m^2/s), in case of lateral inflow q > 0 and k = 0 and for lateral outflow q < 0 and k = 1,
S_f	= energy slope (m/m)
Q	= discharge (m^3/s)

Solution of the equation. For solving this equation, an upstream boundary condition in terms of discharge and a downstream boundary condition in terms of water surface elevation are required.

The equation does not have any analytical solution. In general case, it is discretized in order to obtain a numerical solution. Knowing the upstream discharge and the downstream water elevation, the water surface profile is integrated, step by step starting from the downstream end.

Integrating the equation between sections *i* and *k*, gives:

$$\int_i^k dH + \int_i^k -\frac{qU}{gA}(k-1)dx + \int_i^k S_f dx = 0 \tag{7.2}$$

it gives:

$$H_k - H_i - (k-1)q\frac{\Delta x_{ik}}{2g}\left(\frac{V_k}{A_k} + \frac{V_i}{A_i}\right) + \frac{S_{f_i} + S_{f_k}}{2}\Delta x_{ik} = 0 \qquad (7.3)$$

it can be written as:

$$H_i(Z_{wi}) = H_k + \Delta H(Z_{wi}) \qquad (7.4)$$

Where Z_{wi} = water surface elevation (m).

A subcritical solution exists if:

$$\delta = H_k + \Delta H(Z_{ci}) - H_i(Z_{ci}) > 0 \qquad (7.5)$$

Where Z_{ci} = critical water elevation (m) defined at section i by $Fr_i^2 = 1$.
If $\delta > 0$ then a subcritical solution exists and if $\delta < 0$ then there are supercritical flow conditions in which critical depth is assumed asymptotically. The water surface profile is then overestimated.

7.2.3. Unsteady flow module

The unsteady flow module computes the water surface profile in the canal under unsteady flow conditions. The initial water surface profile is provided by the steady flow module.

Saint-Venant equations. Two equations are needed to describe unsteady flow in open channels: the continuity equation and the momentum equation.

The continuity equation, which accounts for the conservation of the mass of the water is expressed as (Chow, 1959):

$$\frac{\partial A}{\partial t} + \frac{\partial Q}{\partial x} = q \qquad (7.6)$$

The momentum equation or dynamic equation is expressed as:

$$\frac{\partial Q}{\partial t} + \frac{\partial Q^2 / A}{\partial x} + gA\frac{\partial z}{\partial x} = -gAS_f + kqV \qquad (7.7)$$

Where
z	= water elevation (m)
q	= lateral inflow or outflow (m²/s)
k	= 1 for outflow, and k=0 for inflow
V	= velocity of inflow or outflow (m/s)

These partial differential equations are completed by initial and boundary conditions in order to be solved. The boundary conditions are the hydrographs at the upstream nodes of the reaches and a rating curve at the downstream node of the model (because subcritical conditions prevail).

Implicit discretization. Saint Venant equations have no known analytical solutions in real geometry. They are solved numerically by discretizing the equations: the partial derivatives are replaced by finite differences. The discretization chosen in this model is a four-point implicit scheme known as Preissmann's scheme (Cunge et al., 1980).

7.2.4. Sediment module

The sediment module allows the computation of sediment transport in steady and unsteady state flow conditions. It is capable of determining the sediment deposition and scouring along the system, change in bed levels, outlets/offtakes sediment extractions, and sediment affected system's hydraulic performance.

Equilibrium sediment transport (sediment transport capacity)

For computing the sediment transport capacity of the irrigation canals the model uses a number of formulae as Meyer-Peter (1948), Einstein (1950), Bagnold (1966), Engelund-Hansen (1967), Ackers-White (1973) and Van Rijn (1984). The formula which fits best to the actual conditions can be selected for further simulations. Some of these formulae are given in Appendix A.

Non-equilibrium transport

One dimensional sediment transport is classically modelled by the convection diffusion equation:

$$\frac{\partial Ac}{\partial t} + \frac{\partial (AUc)}{\partial x} - \frac{\partial}{\partial x}\left(K_D \frac{\partial c}{\partial x}\right) = \varphi \tag{7.8}$$

Where
A	= flow area (m^2)
c	= sediment concentration (m^3 of sediment/m^3 of water)
t	= time (seconds)
x	= distance (m)
K_D	= diffusion coefficient (m^2/s)
φ	= sediment exchange rate with the bed (kg/m/s)
U	= mean flow velocity (m/s)

Solution under steady state conditions

In steady state the variation with respect to time is not considered, therefore the first term disappears. The dimensional analysis shows that the diffusion term can be neglected in steady state solutions (Baume, 2005). Therefore the above equation simplifies to:

$$\frac{\partial AUc}{\partial t} = \varphi \tag{7.9}$$

Model of exchange term φ. The exchange term represents the flux (mass per unit time per unit length) of the material brought to flow. For solute transport, it stands for adsorption, desorption or any other chemical reaction.

The first order degradation reaction, such as auto-eruption process in rivers is given as:

$$\frac{dc}{dt} = -kc \tag{7.10}$$

where k is the reaction constant. The above equation develops as:

$$\frac{\partial c}{\partial t} + U \frac{\partial c}{\partial x} = -kc \tag{7.11}$$

and corresponding degradation is written as:

$$\varphi = -kAc \tag{7.12}$$

In sediment transport modelling this exchange rate is rather complicated. In uniform canals, this flow rate is supposed to reach a sediment transport capacity Q_s^*. Many formulae have been developed as mentioned under equilibrium conditions to assess this quantity.

If sediment input is higher than the sediment transport capacity of the canal then deposition will take place and if it is less then erosion will take place in the canal. For fine sediments the adaption does not take place immediately, but it takes some distance to reach at equilibrium conditions. This may be represented by the non-equilibrium model as:

$$\varphi = \frac{\partial Q_s}{\partial x} = \frac{1}{L_A}(Q_s^* - Q_s) \tag{7.13}$$

Where
L_A = adaptation length (m)
Q_s^* = equilibrium sediment transport capacity (m³/s)
Q_s = actual sediment discharge in the canal (m³/s)

The expression for L_A can take into account the role of the viscosity, the particle diameter and the turbulence such as Han's (1980) formula:

$$L_A = \alpha \frac{u^*}{w_f} \tag{7.14}$$

Where
α = a calibration parameter

u^* = the shear velocity (m/s)

w_f = fall velocity of the particles (m/s)

The fall velocity can be estimated by the following empirical formula (Baume, 2005):

$$w_f = 10\frac{v}{d}\left[\left(1+\frac{0.01(s-1)gd^3}{v^2}\right)^{0.5}-1\right]$$ (7.15)

In case of more sediment classes the total sediment transport capacity is calculated as a combination of sediment transport capacities, calculated with the representative diameters of each class and weighed by proportion of each class in the total load.

Sediment diversions at junctions. The first principle is the mass conservation at each node of the model. For a convergent system, this principle is sufficient for the complete resolution:

$$c_3 = \frac{Q_1 c_1 + Q_2 c_2}{Q_1 + Q_2}$$ (7.16)

The subscripts 1 and 2 designate the upstream canals and 3 designates the downstream canal.

For divergent systems, where upstream canal 1 splits into two downstream canals 2 and 3, then it is simply assumed that $c_2=c_3=c_1$, which is generally verified as far as solute or very fine particles (clay or fine silt) are concerned, but not for coarser particles. Measurements are needed for coarser particles to determine this distribution. Numerical experiments have shown that the concentration in the offtakes is dependent on the flow and offtake geometry (Baume, 2005). It is calculated in the model as:

$$Q_s^{offtake} = \theta Q_s^{canal}\frac{Q^{offtake}}{Q^{canal}}$$ (7.17)

Where

$Q_s^{offtake}$ = sediment discharge in offtake (m³/s)

Q_s^{canal} = sediment discharge in canal (m³/s)

θ = ration of sediment discharge in offtake to the canal

$Q^{offtake}$ = water discharge in offtake (m³/s)

Q^{canal} = water discharge in canal (m³/s)

The value of θ may be different from 1, particularly for sand and needs to be measured in the field. Belaud and Paquier (2001) have given some typical values of θ, depending upon the sediment and velocity distribution in the cross-section and offtake geometry.

Concentration distribution. Rouse (1937) formula is applied to determine the vertical concentration profile. The formula is given as:

$$\frac{c}{c_a} = \left(\frac{y-z}{z} \frac{a}{y-a} \right)^m \tag{7.18}$$

with:

$$m = \frac{w_f}{\xi \kappa u_*}$$

Where
a	= reference height, limit height between suspended and bed load (m)
z	= given water depth (m)
y	= water depth (m)
c_a	= reference concentration (m³/s)
κ	= Von-Korman constant
U^*	= shear velocity (m/s)
ξ	= ratio between vertical diffusion coefficient and turbulent viscosity.

Practically, the value of reference level is very small, with a/y ≪1, the above equation can be defined as:

$$c(z) = A_c \left(\frac{y-z}{z} \right)^m \tag{7.19}$$

with:

$$A_c = c(a/y)^m$$

The transverse distribution is supposed to be of the same type as the velocity, with a different exponent.

Bed evolution. Bed aggradation or degradation is obtained by conservation equations. The exchange rate represents the material lost by the bed and equals:

$$\varphi = -\rho_s (1 - p_r) \frac{\partial A_b}{\partial t} \tag{7.20}$$

Where
ρ_s	= sediment density (kg/m³)
p_r	= bed porosity
A_b	= bed area (m²)

Computation procedure. In steady and unsteady flow computations, sediment transport is computed at each time step after the hydraulic computations.

Then the sediment transport capacity is calculated in the calculation sections *i*. It depends on the hydraulic and geometric variables and the sediment properties along with adaptation length. In next step the non-equilibrium sediment transport equation is solved, which is:

$$\frac{\partial c}{\partial x} = \frac{1}{L_A}(c^* - c) \tag{7.21}$$

The adaptation length, L_A, and the sediment transport capacity c^* are approximated by their averages between calculation sections *i* and *i*+1. Then the solution becomes:

$$c(x_{i+1}) = c(x_i) + \left(c^* - c(x_i)\right)e^{-\frac{x-x_i}{L_A}} \tag{7.22}$$

In case of several classes the total sediment transport capacity is calculated as:

$$c_i^* = \sum_{j=1}^{N_t} p^{j,i-1} c^*(d^j) \tag{7.23}$$

Where

$p_{j,i}$ = the proportion of sediments in the class j in section *i*
N_t = the number of transported classes
d^j = representative diameter for class j

The adaptation length for each class is calculated as:

$$L_{Ai}^j = \frac{u_{*i}}{\alpha w_f^j} \tag{7.24}$$

Where w_f^j is the fall velocity of class j.

Bed evolution. In section *i* the bed variation during time *t-dt* and time *t* is given by sediment continuity equation as:

$$dA_{bi} = -\frac{dt}{1-p_r}\varphi_i = -\frac{dt}{1-p_r}\left(\frac{Q_i}{L_{A,i}}(c_i^* - c_i) + c_i I_i\right) \tag{7.25}$$

Where I_i is the seepage rate at section *i*.

Solution under unsteady flow

In unsteady flow the same formulae are used as for steady flow. However, the convection diffusion equation is solved in its whole. The Holly-Preissmann scheme is used for the convective part. This algorithm uses the method of characteristics. The diffusion terms are solved by the Crank-Nicholson method. These solutions are described in Appendix C.

Model calibration

Extraction ratios or coefficients of vertical influence can be calibrated with field data. If not available, α can be set to 0.5 and θ can be set to 1 for fine particles. Sediment transport in reaches is adjusted with parameters β (sediment transport parameter) and α (adaptation parameter):

$$C^*_{model} = \beta C^*_{formula} \tag{7.26}$$

$$1/L_A = \alpha f(u^*, w, h, U) \tag{7.27}$$

Parameters β can be adjusted on measured sediment discharge if an equilibrium stage is reached. Integrative calibration consists of adjusting of the parameters β and α so that the model results match observed topographic evolutions. The diffusion coefficient K_D can be adjusted on field data as well.

7.3. Sediment transport model set up

Before using any model, it needs to be evaluated in terms of its strengths and weaknesses, like its sensitivity to various parameters, its robustness to different inputs and its limitation to perform a certain analysis. After that every model needs proper calibration in light of the field measurements on hydraulic and sediment transport. After satisfactory calibration and validation, the model becomes able to simulate various scenarios. It can then predict quite precisely the effects of certain inputs and modifications on the targeted output. Regarding sediment transport modelling in irrigation canals, the canals' geometry, canals' hydraulics and the incoming sediment rate and type are the main influencing factors, the effect of which can be tested by model simulations.

In this chapter all of these factors have been considered and the sediment transport model, SIC has been evaluated, calibrated and then used for scenario simulations. First of all a sensitivity analysis was performed and the effects of certain inputs were observed on the model's output. A number of sediment transport formulae available in the model were evaluated in order to see their robustness to various inputs. Then the formulae showing the stable behaviour were selected for further simulations. The model was calibrated with one set of field data on hydraulics and sediment transport and then validated with the second set of field data. Then it was used for various water management and sediment transport scenario simulations.

7.3.1. Sensitivity analysis

A sensitivity analysis was performed in order to see the effects of various inputs, which were water flow, sediment inflow, sediment particle size, etc. on the model's output, which was sediment deposition volume in this case. The canals downstream of RD 242, the Lower Machai Branch Canal and the Maira Branch Canal were simulated at the design discharge with a sediment concentration of 1.0 kg/m³, with mean particle diameter of 0.10 mm. The parameters like water inflow, mean sediment inflow, median diameter and canal roughness were changed and the effects were observed on the sediment deposition volume in the canal. Simulations were performed for a period of one year and the Engelund-Hansen (1967) predictor was used.

The results of the sensitivity analysis are given in Table 7.1. It can be seen that the most influential parameter was the rate of sediment inflow, whereas the least influential parameter was the canal roughness coefficient. The variation in the rate of water inflow is a moderate influent. The variation in initial concentration does not affect much.

Table 7.1. Results of sensitivity analysis

Parameter	Initial value	Modified value	% modification	Initial deposition (m³)	Modified deposition (m³)	% change
Q (m³/s)	29.50	26.60	-10	58,600	49,300	-15.74
c (kg/m³)	0.800	0.720	-10	58,600	46,000	-21.44
Manning's n (s/m^{1/3})	0.023	0.021	-9	58,600	57,500	-1.86
d_{50} (mm)	0.177	0.153	-14	58,600	58,000	-1.01
Initial c (kg/m³)	0.700	0.630	-10	58,600	58,600	-0.01

7.3.2. Comparison between sediment transport predictors

Almost all of the sediment transport predictors have been developed under a controlled environment in the hydraulic laboratories with quite simplified conditions using particular particle size ranges. Majority of these predictors has been developed for coarser particles and a few of them have been developed for finer particles. Usually these predictors give good predictions for the hydraulic and sediment transport conditions under which they were developed. A universal sediment transport predictor which can be applied for every hydraulic and sediment transport condition is not available. Therefore it becomes extremely important to know the strengths and limitations of the sediment transport predictors before their use. Therefore a number of three mostly used equilibrium sediment transport predictors were compared by simulating sediment transport in automatically downstream controlled irrigation canals.

These predictors were tested on two different flow conditions: one on actual water flow conditions and the other on design water flow conditions, in order to see the effect of changes in velocity and flow on the sediment transport predictions. The reason to select these two conditions was operation of the irrigation canals under study. These irrigation canals were operated at lower discharges than the design discharges due to different reasons like sediment deposition in the canals, low crop water requirements and bad canal maintenance, which can be said as existing discharges. The other condition was the canal operation at design discharge, which is usually desired in IBIS irrigation canals. The sediment deposition volume and deposition pattern were compared in this analysis in order to evaluate the predictors.

The sediment transport formulae of Bagnold (1966), Engelund-Hansen (1967) and Van Rijn (1984) were tested. The mean sediment inflow to the canal with actual d_{50} was used in these simulations, which is given in Table 7.2. It can be seen that having the same hydraulic conditions, same sediment inflow with same d_{50}, different formulae have different predictions.

Table 7.2. Description of water and sediment flow parameters for predictor comparison

Flow condition	Description of parameter	Quantity of parameter	Units
At actual flow	Flow from Machai Branch	7.5	m³/s
	Concentration at RD 242	0.140	kg/m³
	d_{50}	0.050	mm
	Contribution in flow from PHLC	14.9	m³/s
	Concentration from PHLC	0	kg/m³
	Total water flow	22.4	m³/s
	Total concentration	0.047	kg/m³
At design flow	Flow from Machai Branch	7.5	m³/s
	Concentration at RD 242	0.140	kg/m³
	d_{50}	0.050	mm
	Contribution in flow from PHLC	20.7	m³/s
	Concentration from PHLC	0	kg/m³
	Total water flow	28.2	m³/s
	Total concentration	0.037	kg/m³

Sediment deposition volumes in the canals are given in Table 7.3. It shows that at actual flows (or low flow) the difference in deposition volume was not too much but at design flow (or high flow) the difference in predictions was too large. It means that some of the formulae are quite sensitive to flow velocity or other hydraulic parameters. It can be seen that Engelund-Hansen (1967) formula showed quite reasonable variations -16% in terms of its comparison at low and high water discharges. The Van Rijn (1984) and Bagnold (1966) formulae showed somehow big variations -54% and -40%, which showed that these formulae may be quite sensitive to high flow velocities.

It was found that there was not much difference in the sediment deposition volumes, but there were quite big variations in canal bed evolution. Engelund-Hansen formula showed deposition in the head reaches whereas at the same time other formulae resulted in erosion in those canal reaches. Figure 7.1 presents difference in simulated bed levels in comparison with initial bed levels. At the last reach Bagnold and Engelund-Hansen formulae showed deposition, whereas Van Rijn formula showed erosion trend. These discrepancies in the results of various predictors can be attributed with the environment under which these relationships were developed. This means during simulations it depends where the hydraulic conditions in the canals come in the range of the hydraulic parameters under which the particular formula was developed. So it gives better results there. It becomes then increasingly important to know that which formula(e) fit(s) in which particular canal flow and sediment conditions in order to predict the true simulated values.

Table 7.3. Results of formulae comparison

Predictor	Dep. vol. at actual flow (m^3)	Dep. vol. at des. flow (m^3)	Difference %
Engelund Hansen	20,000	17,000	-16
Bagnold	18,500	11,000	-40
Van Rijn	17,800	8,000	-54

On the basis of these simulations it can be said that the Engelund-Hansen (1967) predictor is more robust and reliable for simulating sediment transport under different hydraulic conditions. However, this formula can only be used to simulate sediment transport in irrigation canals after properly calibration and validation according to the field conditions of water flow and sediment transport.

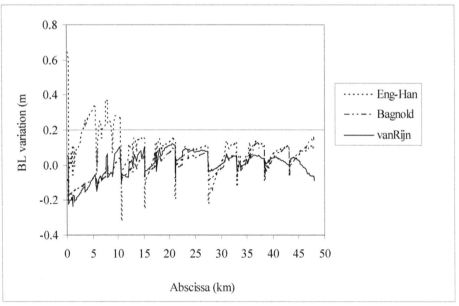

Figure 7.1. Simulated bed levels under actual water flow and sediment concentration

7.3.3. Model calibration and validation

In sediment transport modelling, two types of calibrations can be considered one is instantaneous approach and other is integrative approach. In instantaneous calibration the parameters of the formulae are calibrated with the instantaneous data. In case of sediment transport, it can be the evolution formulae for the concentrations using the actual concentrations at different abscissa but at the same time. In integrative calibration the parameters of the formulae (which deal with concentration) in relation to topographic data are considered. This calibration is related to a period of bed evolution. It consists in minimizing the difference between two geometries, one obtained by simulation from initial time to final time of simulation and the other is the measured geometry at final time (Belaud, 1996).

Vabre (1995) used both of these methods and found that integrative method is more reliable. The variability in time for the concentration can be very high (due to physical variations as well as measurements accuracy), but the variations of the volumes are supposed to integrate this variability. Even from irrigation management and maintenance point of view the difference in bed levels and the volumes of sediment to be dredged out are more important than determining the sediment concentrations along the canal.

In this study the integration approach has been used to calibrate the model. The initial and final simulated bed levels were compared with the measured bed levels. The model was calibrated by adjustment of three factors namely the sediment transport predictor factor β, 1.0 for the original formula, adjustment of adaptation length by adjusting the deposition coefficient, α_d and erosion coefficient, α_e. Various combinations of α_d and α_e were utilized and the best value of $R^2 = 0.67$ was found at $\alpha_d = 0.002$ and $\alpha_e = 0.008$. After calibration the model was validated by using cross sectional survey data of July 2008. The sediment inflow data for the year 2008 are given in Table 7.4.

Table 7.4. Flow and sediment data used in simulations

Day	Sediment inflow at RD 242 kg/m^3	Discharge at RD 242 m^3/s	Discharge at confluence m^3/s
0	0.097	6.2	16.4
30	0.097	6.5	16.4
60	0.149	7.2	20.3
90	0.306	7.8	19.4
120	0.322	8.2	16.7
150	0.163	8.4	18.9
180	0.109	7.6	19.6
210	0.072	6.8	20.5
270	0.059	6.2	19.9
300	0.059	6.2	18.1

The simulation was performed for 120 days from April 2008 to July 2008 because the canal operation was started in April 2008. Until this period the canal was kept closed, in January 2008 for routine maintenance and then for further two months due to delay in maintenance activities due to different reasons. During field measurements the cross-sections were measured less frequently all along the canal because cross sectional survey in flowing water is a quite difficult and time consuming exercise. The model validation results are given in Table 7.5.

Table 7.5. Values of calibration parameters

S. No	Parameter Description	Calibration values	Validation values
1	ME (m)	0.264	0.358
2	RMSE	0.094	0.103
3	EF	0.746	0.646
4	CRM	-0.003	-0.004
5	MAE (%)	-0.090	-0.110

7.4.   Scenario simulation

Field measurements are very important for the assessment of the sediment transport in any irrigation canal. But the collection of field data for sediment transport is a very cumbersome and time consuming process. Whereas to determine the behaviour of water and sediment transport at different canal operations is quite difficult due to resistance from the water users to any change in canal flows. In this situation, the field measurements can be conducted on the available water and sediment flow conditions in the canal. On the basis of these measurements, a sediment transport model can be set up and then used for scenario simulations in order to see the effect of a variety of scenarios of canal operation and sediment concentration on it operation and maintenance.

The imminent issue, in the lower part of the system under study, in the PHLC, the Lower Machai Branch Canal and the Maira Branch Canal is the increased sediment inflow from the Tarbela Reservoir. In Chapter 4 the likelihood of increased sediment discharge from the reservoir to these irrigation canals has been discussed. A moderate to severe quantity of sediment discharge is likely to be discharged into these irrigation canals.

Further, the downstream controlled irrigation canals are not being operated as per their conceived operation strategy. It was the canal operation under CBIO (see Chapter 5) in order to control waterlogging in the command area and to avoid sedimentation in the canals. The CBIO was not fully implemented because of water users resistance to canal flow reductions and secondary canal closures. In sediment transport modelling various other options of canal operations have been simulated in order to help in the decision making process on water and sediment management. A variety of such scenarios and their likely effects on canals' efficiency and maintenance are discussed in this chapter.

The main scenarios, which are discussed in this chapter start with the simulation of actual discharge (or existing discharge) with actual sediment concentration and then the increased (or designed) discharge with actual sediment concentration. The proposed canal operation strategy, conceived during the canal design, the CBIO has also been simulated with the actual inflow of sediment and with the increased inflow of sediment. In CBIO a group of the secondary canals are closed for a certain period, depending upon CWR, (see Chapter 5) in order to match irrigation water supply and crop water requirements. The effect of the number of grouping options for the closure of secondary offtakes also has been assessed on the sediment transport in the canal downstream of RD 242.

The effect of various scenarios of sediment transport on hydraulic performance of the canal has been assessed. The question how spatio-temporal sediment deposition affects the flow releases from automatic discharge controllers has been investigated. Then in a broad context, how the sediment transport affects the self regulating structures in the canal and how the changes in bed levels affect the response of such water level control structures.

7.4.1. Simulations under existing conditions of water and sediment discharge

Figure 7.2 presents simulated initial and final bed levels (BL). Simulations were performed for a period of one year. The sediment data is given in Table 7.5, where the median sediment size was in the range of coarse silt and ranged from 0.050 to 0.065 mm. The maximum sedimentation took place about 0.40 m to less than 0.2 m in the head reach, whereas in rest of the canal the deposition depth was less than 0.2 m. In the middle of the canal some erosion took place. The simulated deposition volume was about 19,400 m^3 in a one year period. The average flow at the confluence remained 18.6 m^3/s.

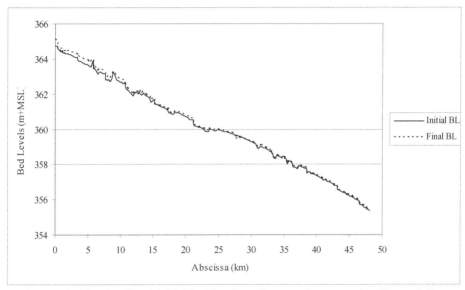

Figure 7.2. Simulated and initial bed levels under actual water and sediment inflow downstream of RD 242

7.4.2. Sediment transport under design discharges with existing sediment concentration

As the actual discharges in the canal were less than the design discharge. Simulations were performed in order to see how sediment would behave if the canal was run at design discharge. The sediment quantity was kept same as in the previous simulation. The simulation results are presented in the Figure 7.3. If the bed levels are compared with the bed levels under actual discharge conditions, it can be seen the trend is more or less same except at the head reach, particularly at the confluence where scouring took place, instead of deposition. The sediment deposition volume under these conditions was computed to be 18,000 m³ in a one year period of canal operation. The design flow used in this simulation was 25.2 m³/s.

7.4.3. Sediment transport under CBIO with existing sediment concentration

One more scenario was simulated in order to see the effect of canal operation as per CBIO, on sediment transport. The flow hydrograph under CBIO is given in Figure 5.2, whereas sediment data has been given in Table 7.4. The variation in canal bed level is given in Figure 7.4. It can be seen that the raise in bed levels under CBIO was comparatively higher than the sediment deposition under design and actual canal operations. It is because at some times the canal was operated at almost 50% of the full supply discharge, which caused ponding effects in the canal and dropped the sediment transport capacity quite low. The phenomenon resulted in more sedimentation. The deposited volume was computed to be 22,400 m³ under these operations. The sediment deposition pattern under existing flow conditions and under CBIO is almost the same. The existing sediment inflow mainly consisted of coarse silt, hence it was mainly deposited in the head reaches.

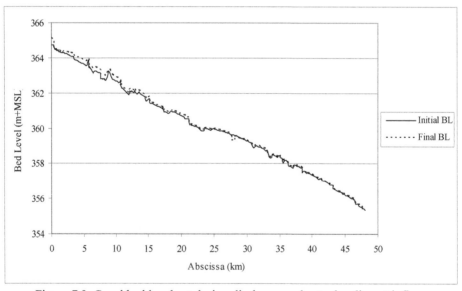

Figure 7.3. Canal bed levels at design discharge and actual sediment inflow.

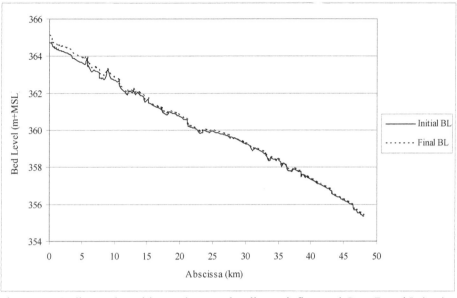

Figure 7.4. Sediment deposition under actual sediment inflow and Crop Based Irrigation Operations

7.5. Sediment transport in PHLC

Till now, only sediment transport in the system downstream of RD 242, the Lower Machai Branch Canal and the Maira Branch Canal, has been discussed. In this section sediment transport in PHLC will be highlighted. PHLC can be seen as a parent canal of Maira Branch Canal as it carries water from the Tarbela Reservoir to the Maira Branch Canal (Figure 5.1). It also supplies water to its own four secondary offtakes and a number of 24 direct outlets.

At this stage it is quite difficult to say that how much will be the sediment discharge from the Tarbela Reservoir due to non-availability of particular studies on this topic. A review of the general available studies on sediment transport in the Tarbela Reservoir was given Chapter 4. On the basis of this review a firsthand estimate can be made on how much sediment will be discharged from the reservoir and what would be its temporal variability along with the characteristics of the discharged sediment. In a very simple case that it can be assumed that the sediment discharge to the PHLC would start from fine particles with low concentration and would gradually move towards the coarser particles with higher concentrations. Anyhow, owing to the sediment transport capacity of the irrigation canals, this study has been limited to sand (fine sand) and silt (coarse to fine silt).

In sediment transport modelling in PHLC, first of all the sediment transport capacity of the canal has been assessed under three different conditions of canal operation, full supply discharge, existing discharge and the CBIO with fine silt to fine sand sediment sizes.

7.5.1. Sediment transport capacity in PHLC

Determination of sediment transport capacity gives first hand information that how much sediment can be carried by the flow under different circumstances. Therefore it becomes crucial to know the sediment transport capacity of the canal in order to have an idea about the behaviour against different sediment inflow and canal operation conditions. A variety of parameters affects the sediment transport capacity, amongst which the discharge in the canal and the sediment size are the most influencing parameters. Further, these parameters are quite fluctuating in the demand based systems. The flows in the canal keep on fluctuating due to changes in crop water demands. The sediment sizes also change in different seasons in a year, particularly they are quite variable during rainy seasons. Therefore the sediment transport capacities of the PHLC and Maira Branch Canals have been assessed under different discharges and sediment sizes in order to know their effect on canals' operation and maintenance.

Figure 7.5.A presents the sediment transport capacity of PHLC at full supply discharge. Keeping in view the sediment discharge from the Tarbela Reservoir, four median particle sizes have been used for this purpose. These median particle sizes are fine sand (0.09 mm), coarse silt (0.044), medium silt (0.022) and fine silt (0.011). The geometric mean sizes of these four sediment classes have been adopted from Alluvial Channel Observation Project (ACOP, 1985). For fine sand the sediment transport capacity ranges from 1.00 kg/m^3 to 0.7 kg/m^3 along the canal, whereas for coarse silt it ranges from 2.00 to 1.00 kg/m^3 and for medium silt it's range is 1.5 to 4.0 kg/m^3. The sediment transport capacity is generally higher downstream of the cross regulators and lower at the upstream of the cross regulators.

Under CBIO the canal supplies are usually kept at full supply discharge (FSD), 80% of the FSD, 75%, 67% and then 50% of the FSD. Whereas full supply discharge and the 50% of the full supply discharge are the two maximum and minimum discharges available in the canal under CBIO. Therefore the sediment transport capacity of each sediment size has been determined. The sediment transport capacity under 50% of the full supply discharge is given in Figure 7.5.B. It was observed that the sediment transport capacity was much lower under 50% discharge conditions than the sediment transport capacity under 100% discharge. For instance sediment transport capacity for fine sand under 50% flow, ranges from less than 0.1 to 0.15 kg/m^3 and similarly for coarse silt it ranges from 0.15 to 0.25 kg/m^3 at different locations along the canal.

The Figures 7.5.A and 7.5.B show that the sediment transport capacity decreases as the median particle size increases. As a reduction of only 40% in the flow, sediment transport capacity reduces from 5 to 10 times at different locations along the canal. This decrease in sediment transport capacity can be attributed to the reduction in flow velocities and ponding effect in the downstream controlled canals. Sediment transport capacity through the siphons is not truly representative of the siphons because siphons were modified during the modelling due the limitation in the model. The siphons in the canal were replaced by flumes with steep slopes corresponding to the drop in the energy line at the starting point and end point of the siphons.

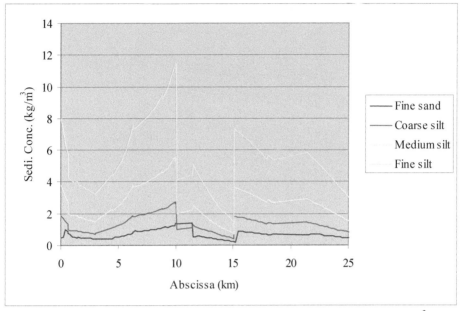

Figure 7.5.A. Sediment transport capacity at full supply discharge (Q = 27 m³/s)

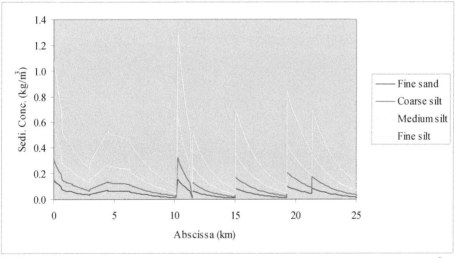

Figure 7.5.B. Sediment transport capacity at 50% of full supply discharge (Q = 16 m³/s)

7.5.2. Sediment transport capacity in the canals downstream of RD 242

Similarly, Figures 7.6.A and 7.6.B present the sediment transport capacity of the canals downstream (d/s) of RD 242 for the above mentioned sediment sizes and the two flow conditions of full supply discharge and the 50% of the full supply discharge. Sediment transport capacity for fine sand is less that 0.2 kg/m³ all along the Maira

Branch Canal, whereas for coarse silt it ranges from 0.1 to 0.4 kg/m^3 along the canal at different locations.

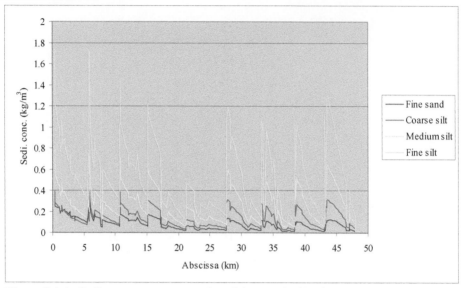

Figure 7.6.A. Sediment transport capacity of the canals d/s of RD 242 under full supply discharge

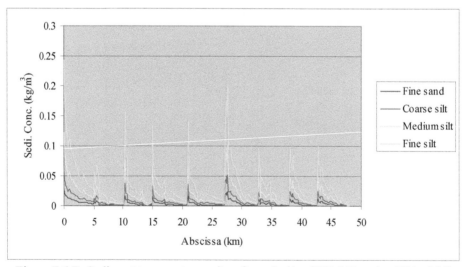

Figure 7.6.B. Sediment transport capacity of canals d/s of RD 242 under 50% of full supply discharge

Similar trend has been observed in the canals downstream of RD 242 as a reduction in sediment transport capacity with the decrease in water discharge. Sediment transport capacity in the Maira Branch Canal is quite low as compared to the PHLC. Under 50%

flow conditions or under half of the full supply discharge the sediment transport capacities drop a lot. For fine sand the sediment transport capacity is not more that 0.02 kg/m^3 at all the locations along the canal and for coarse silt it ranges from 0.01 to 0.05 kg/m^3 at different locations along the canal.

7.6. Scenario simulations with Tarbela effect

In order to see the effects of sediment inflow from the Tarbela Reservoir, a variety of scenarios of sediment inflow with canal operations were generated with varying sediment concentration and particle sizes. These scenarios have been developed on the basis of sediment inflow into the Tarbela Reservoir, sedimentation in the reservoir, dam operations and sediment transport in PHLC and operation of the PHLC and Maira Branch Canal.

The Surface Water Hydrology Project report (2008) states that in the Indus River systems, about 95% of the total sediment comes in summer and monsoon season (June to September) in Indus River at rim station upstream of Tarbela Reservoir. Therefore more sediment discharge into the canal can be expected in these months, after filling of the reservoir with sediment. Retaining the pool level in the reservoir also affects sediment discharge from the reservoir to the PHLC. As up to September every year, the maximum pool level, 470 m+MSL is maintained, depending upon water availability in the upper river basin and the demand in the downstream river and irrigation system. After every September the pool level starts to decline and it drops to 440 m+MSL. The decline in pool level continues till the end of April or start of May. From February to April pool level further drops up to a minimum of 420 m+MSL. This drop in pool level causes reworking of the delta and brings sediment into the offtaking tunnels. So sediment discharge from the reservoir can also be expected in these four months from February to May. These will be the two prominent reasons of sediment discharge into the offtaking tunnels. Then gradually, the sediment discharge to the tunnels will increase as the sedimentation takes place in the reservoir.

On the other hand, the PHLC and Maira Branch Canals are fully demand based and semi-demand based irrigation canals, respectively. A small fraction of the flow from PHLC is withdrawn on the fully demand basis, which is less than 5% of the full supply discharge of PHLC. Hence, it can be said that the PHLC also flows on the semi-demand based operations, on the principles of CBIO. Both of the canals are generally run for eleven months in a year and remain closed for one month, during January every year, for repair and maintenance works. Therefore, the canal flows for about 330 days per year.

Keeping in view the above factors regarding sediment inflow to the reservoir, sedimentation in the reservoir and the dam and canal operations, the following scenarios have been generated. For simplicity, the sediment discharge from the dam has been divided into three periods, based on the sediment concentration and discharge in the Indus River: Period-I, from February to May; Period-II from June to September; and Period-III from October to December. The maximum sediment in the canal can be expected during Period-II, June to September as more that 95% of the total sediment comes into the river in these four months. Then, also high sediment discharge can come during Period-I, February to May, as pool level drops in these months. A drop in pool level after a certain point causes reworking of the delta, which causes high sediment concentration in the reservoir flow due to erosion of the deposited material. Then a quite small amount of sediment may come in Period-III, September to December. As after filling of the reservoir, quite clear water flows in the reservoir. So a very small amount of sediment would come in these three months.

Fairly fine particles are expected in the sediment discharge from the dam. As coarser particles are settled quite upstream of the reservoir, where deceleration of the flow takes place and only fine particles travel along the flow, which continue settling along the flow path, depending upon the hydro-dynamic conditions in the reservoir. So, sediments ranging from fine silt to fine sand have been considered for simulations. Very fine clay particles usually do not settle in the canals even at 50% of the full supply discharge conditions. Therefore simulations have been limited to fine silt. The PHLC tunnel intake level is about 20 m higher than the other main four tunnels, so comparatively less sediment discharge can be expected into the PHLC. Table 7.6 gives information on sediment discharge from the reservoir into PHLC and the data used for sediment transport simulations in the PHLC and Maira Branch canals.

Table 7.6. Criteria for sediment scenario preparations

Parameter	Description	Period-I Feb-May	Period-II Jun-Sep	Period-III Oct-Dec
Concentration (kg/m³)	Minimum	0.025	0.100	0.005
	Medium	0.125	0.500	0.025
	Maximum	0.250	1.000	0.050
Median particle size (mm)	Fine silt	0.011		
	Medium Silt	0.022		
	Coarse silt	0.044		
	Fine sand	0.090		
Flow conditions (m³/s)	Design Q	27		
	Existing Q	18		
	CBIO	15-27		

7.6.1. Sediment transport at full supply discharge

As mentioned above simulations at full supply discharge were performed for 330 days. It is important to mention that the time period of 330 days represents a complete cycle of one year canal operation because canals in IBIS remain closed for one month in January. In this scenario all the offtaking canals remain at their design discharges. The variations in the PHLC bed levels are given in the Figure 7.7. It can be seen from Figure 7.7 that under such estimation of sediment discharge from the reservoir, the PHLC canal can run for a full year period for fine sand to medium silt. In PHLC sedimentation took place along the whole length of the canal. The more susceptible reach was the reach between Cross Regulator No. 2 (abscissa 11,500) to Cross Regulator No. 3 (abscissa 15,000). As the sill level of Cross Regulator No. 3 is about 0.5 m higher than the canal bed level, which cause lot of sedimentation upstream of the Cross-Regulator No. 3. Similarly sedimentation took place between Cross Regulator No. 4 (abscissa 21,300) and Cross

Regulator No. 5 (abscissa 24,300). Again, the sill level of the Cross Regular No. 5 is 0.3 m higher than the canal bed level. Therefore sedimentation took place upstream of this point. The total sediment deposition volume in PHLC is given in Table 7.7.

Figure 7.8 shows deposition downstream of RD 242 under above given conditions of flow and sediment discharge. Figure 7.8 shows that there was not much deposition of sediment downstream of RD 242, closer to the confluence even for find sand and coarse silt. Downstream of RD 242 the deposition mainly took place in the head reaches for fine sand, in head and middle reaches for coarse silt and all along the canal for medium silt. The total sediment deposition volume in PHLC and downstream of RD 242 with different sediment sizes under design discharge conditions is given in Table 7.7. From Table 7.7 it can be seen that the deposition of fine sand took place mainly in PHLC as well as in the canals d/s of RD 242. The volume of deposition of fine sand downstream of RD 242 is less than the deposition volume in PHLC, which shows that the sediment transport capacity of PHLC is not enough to transport fine sand more than 1.0 kg/m^3. On the other hand the coarse silt and medium silt did not deposit much in PHLC and most of it reached downstream of RD 242.

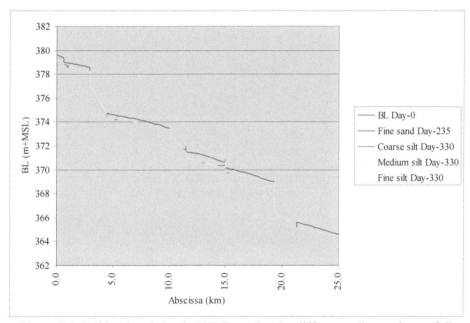

Figure 7.7. Bed level variation in PHLC canal under different sediment sizes at full supply discharge conditions

Table 7.7. Sediment deposition volume in PHLC and d/s of RD 242 under design discharge

Canal	Fine sand m^3	Coarse silt m^3	Medium silt m^3
PHLC	30,500	9,300	2,000
d/s of RD 242	36,700	34,100	8,100
Total	67,200	43,400	10,100

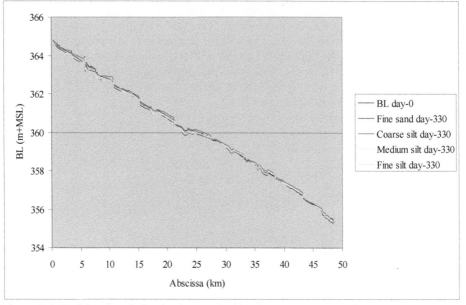

Figure 7.8. Bed level variation in the canals d/s of RD 242 under different sediment sizes at full supply discharge

7.6.2. Sediment transport at existing discharge conditions

Existing flows are less than the full supply discharge due to a number of canal operation and maintenance reasons. As, some of the secondary canals are operated at less discharges due to poor maintenance and are also supplied with less discharge when farmers do not need water. Figure 7.9 presents the bed level variations in PHLC under existing flow conditions. Under these flow conditions sediment transport capacity was less than the sediment transport capacity at full supply discharge. Hence a raise in bed levels was observed earlier, as compared to the bed levels raise under design discharges. It reduced the flow carrying capacity of the canal and caused a raise in water level at the PHLC. The raise in water levels ultimately reduced the flow entering into the canal from the automatic flow control system. In this case the flow started to reduce just after 130 days of operation. Table 7.8 presents total sediment deposition volume in PHLC and d/s of RD 242.

Under these operations more sedimentation took place in PHLC than the deposition under full supply discharge (FSD). Here fine sand deposited much faster as compared to the deposition under FSD. After 188 days of canal operation the simulation stopped because the deliveries from the system head were reduced to so low that they could not meet the discharges of the offtakes. Therefore the simulations were stopped. About 20,000 m^3 and 2,000 m^3 more sediment deposition took place in PHLC under coarse silt and medium silt respectively. It shows that any reduction in flow causes more sedimentation in the canals. These variations show that by reducing flows the hydrodynamic forces required to transport the sediments also decreased and sediment started to drop in the PHLC.

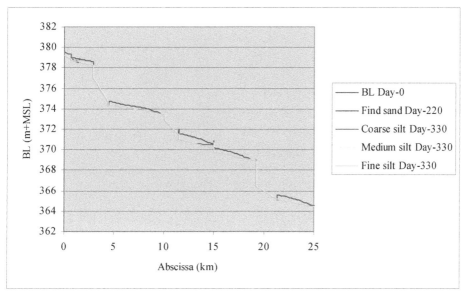

Figure 7.9. Sediment deposition under existing flow conditions in PHLC

Table 7.8. Sediment deposition volume in PHLC and d/s of RD 242 under existing flow conditions

Canal	Fine sand (for 188 days) m^3	Coarse silt m^3	Medium silt m^3
PHLC	43,000	28,700	4,100
d/s of RD 242	39,700	47,100	13,200
Total	82,700	75,800	20,300

Sediment deposition d/s of RD 242 under existing flow conditions is shown in Figure 7.10. Under existing flow conditions, high amounts of sand deposited in PHLC and total sediment deposition in PHLC was higher than the sediment deposition under design conditions. Similarly deposited volumes of coarse silt and medium silt were also higher in Maira Branch Canal than the deposition under full supply discharge.

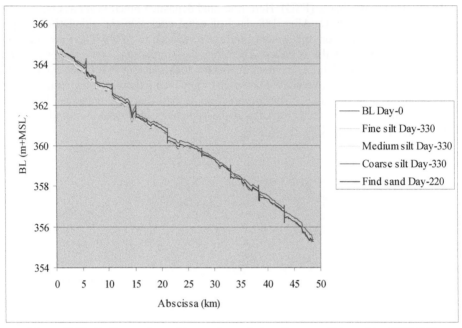

Figure 7.10. Sediment deposition under existing flow conditions d/s of RD 242

7.6.3. Sediment transport under CBIO

The canals under study were recently remodelled and water allowance was doubled in order to meet the peak crop water demands. To avoid the negative impacts of high water allowance like waterlogging and water wastage a canal operation plan the Crop Based Irrigation Operations (CBIO) was devised and implemented in the irrigation canals under study. As the crop water requirements do not remain same throughout the cropping season therefore the water deliveries under CBIO also remain changing in order to meet the crop water demands as shown in Figure 5.2. The changes in flow conditions affect the sediment transport capacity of the canals significantly. Therefore to see the effects of flow variations on the sediment transport in irrigation canals the simulations were performed under CBIO.

The results of sediment transport simulations under CBIO are presented in Figure 7.11 for PHLC. It can be seen from Figure 7.11 that the canal operation under CBIO did not support transport of fine sand. In this case the fine sand mainly deposited in the head reach of PHLC. The deposition in the head reach caused a raise in water levels, which ultimately reduced the flow deliveries from the automatic flow control system. Therefore after 188 days the simulations were terminated because the outflow from the canal was higher than the inflow. The secondary offtakes were then not able to get their design share of water. The other sediments like coarse silt and medium silt were also deposited more under CBIO than under full supply discharge condition and under existing flow conditions.

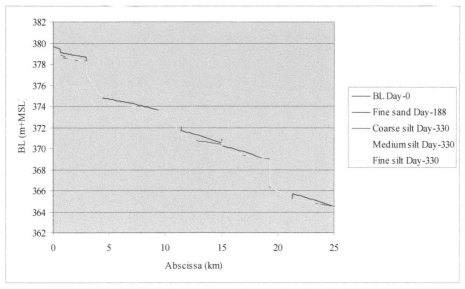

Figure 7.11. Sediment deposition under CBIO in PHLC

Figure 7.12 presents sediment deposition d/s of RD 242 under CBIO. The volume of fine sand deposited under CBIO was less than the volumes deposited under full supply discharge and under existing discharge conditions. It is because of the more fine sand deposition in PHLC. For fine sand the simulations were stopped after 188 days. It is because of the limitations in the model, that when outflow from the canal increases above the inflow, the simulations are terminated.

The total sediment deposition volumes for fine sand, coarse silt and medium silt are given in Table 7.9. The second reason of termination of the simulations might be an error in automatic structure behaviour. A reduction in water supply to the canal cause low water levels in the canals and the AVIS/AVIO cross regulators cannot maintain the desired water levels in the canal. Hence the canal then started to flow without any control and this may introduce some numerical instability in the model. Therefore the simulations stopped after 188 days in case of find sand. The volumes of coarse silt and medium silt were also higher under CBIO than under the other two flow conditions. On overall the deposition of coarse silt was about 10% and 16% higher than the sediment deposition under existing flow conditions and under design discharges respectively. These results show that the canal operations under CBIO cause relatively more sedimentation in the canal.

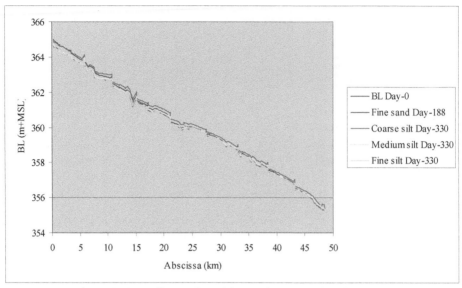

Figure 7.12. Sediment deposition under CBIO d/s of RD 242

Table 7.9. Sediment deposition volume in PHLC and d/s of RD 242 under CBIO

Canal	Fine sand (for 188 days) m^3	Coarse silt m^3	Medium silt m^3
PHLC	38,300	33,700	4,900
d/s of RD 242	32,900	51,200	17,000
Total	71,200	84,900	21,900

7.7.   Effect of flow control on sediment transport

Flow control and discharge variation have large effects on the sediment transport capacity of the canal. Table 7.10 presents the effects of discharge variation on the sediment transport capacity for different sediment sizes. A small reduction in flow reduces the sediment transport capacity up to the large extent in downstream controlled irrigation canals. It has been estimated that a 20% reduction in flow reduces about 50% sediment transport capacity of the flow for fine sand. Similarly, a reduction of 50% in flow reduced about 85% sediment transport capacity of the flow for fine sand. Under CBIO the canal is operated under four different discharges, therefore it has been analyzed how the discharge variation affected sediment transport under these conditions.

The reduction in sediment transport capacity is mainly associated with the ponding effect in the downstream controlled irrigation canals. When water withdrawal from the canal is reduced the water starts to store in the canal in the wedge storage. This storage creates ponding effect in the canal. The flow velocities drop quite low and hence the sediment transport carrying capacity of the flow also reduced.

Table 7.10. Discharge variation and reduction in sediment transport capacity in the downstream controlled irrigation canals

Sediment class	Reduction in sediment transport capacity			
Reduction in flow (%of FSD)	80%	75%	67%	50%
Fine sand (%)	49	75	81	85
Coarse silt (%)	44	66	76	81
Medium silt (%)	41	60	73	79
Average (%)	45	67	77	82

7.8.   Sediment concentration arriving at RD 242 from the PHLC

Though it is quite difficult to predict how much sediment would arrive at without proper calibration of the PHLC sediment transport model, because during the study period the PHLC was a sediment free canal. Anyhow, to have an estimate, the sediment transport simulations were performed for PHLC and following results were obtained. Table 7.11 presents the fraction of different sediment class reaching at the confluence after entering into the PHLC at its head, the Gandaf Outlet.

Table 7.11. Sediment concentration arriving at confluence point from PHLC head (the Gandaf Outlet) at different discharges under CBIO

	Days	Concentration at PHLC head (kg/m^3)	Full supply discharge (m^3/s)	CBIO discharges (m^3/s)			
Discharge (m^3/s)			27.0	23.1	22.1	20.2	18.4
% of FSD			100%	80%	70%	67%	50%
Sand	0	0.25	0.23	0.20	0.15	0.13	0.11
	120	1.00	0.70	0.36	0.18	0.13	0.11
	240	0.05	0.05	0.05	0.05	0.05	0.05
Coarse Silt	0	0.25	0.24	0.20	0.20	0.20	0.17
	120	1	0.90	0.50	0.31	0.22	0.17
	240	0.05	0.05	0.05	0.05	0.05	0.05
Medium Silt	0	0.25	0.25	0.20	0.20	0.20	0.20
	120	1	0.95	0.56	0.38	0.26	0.20
	240	0.05	0.05	0.03	0.02	0.01	0.01
Fine Silt	0	0.25	0.25	0.20	0.20	0.20	0.20
	120	1	0.95	0.69	0.49	0.36	0.29
	240	0.05	0.05	0.05	0.05	0.05	0.05

At the confluence point, sediment also comes from Machai Branch Canal. At this point, flow and sediments from both sources combine and give a new concentration. Table 7.12 presents the combined sediment concentration at the confluence under existing conditions of the flow. As silt is the dominant class from both Machai Branch Canal and PHLC, therefore, the combined sediment concentration is given for coarse silt, with d_{50} equal to 0.044 mm. The concentration at confluence has been estimated on the basis of sediment concentration in the year 2007 from Machai Branch Canal and simulated concentration from PHLC.

Table 7.12. Combined sediment concentration at confluence for existing discharge

	Total flow at confluence	Flow from Machai Branch Canal	Concentration at confluence from Machai Branch	Flow from PHLC	Concen-tration from PHLC	Combined conce-ntration
	(m^3/s)	(m^3/s)	(kg/m^3)	(m^3/s)	(kg/m^3)	(kg/m^3)
Feb-07	26.6	6.00	0.197	20.60	0.225	0.219
Mar-07	26.6	6.00	0.197	20.60	0.225	0.219
Apr-07	23.18	6.50	0.197	16.68	0.225	0.217
May-07	23.18	6.50	0.249	16.68	0.225	0.232
Jun-07	23.18	7.70	0.271	15.48	0.85	0.658
Jul-07	23.18	7.95	0.396	15.23	0.85	0.694
Aug-07	23.18	8.01	0.389	15.17	0.85	0.691
Sep-07	23.18	8.20	0.253	14.98	0.85	0.639
Oct-07	23.18	6.50	0.189	16.68	0.85	0.665
Nov-07	23.18	6.50	0.172	16.68	0.85	0.660
Dec-07	23.18	6.20	0.149	16.98	0.85	0.663

7.9.   Combined sediment concentration at confluence for CBIO

As the sediment transport under CBIO is quite different from the sediment transport under design conditions. The sediment entry from PHLC is also different under CBIO from the existing discharge conditions. After estimating the sediment transport under different discharges and having the hydrograph at the confluence, the combined sediment concentration for coarse silt has been estimated at the confluence point under CBIO and is given in Table 7.13.

The sediment concentration has been estimated for two situations, with minimum and maximum contribution of flow from PHLC. The water availability in the Machai Branch Canal is dependent upon the water availability in Swat River. Therefore in case of less water from the source the Machai Branch Canal feeds it own secondary offtakes and supplies less water to the canal downstream of RD 242 the Lower Machai Branch Canal and the Maira Branch Canal. In case of enough flow from the river, the Machai Branch Canal supplies maximum water to Lower Machai Branch Canal, which is 8.5 m^3/s, whereas in case of minimum flow it can be zero at some times. The maximum flow from PHLC can be 25 m^3/s and depending upon the requirement in the command and contributions from Machai Branch Canal the flow is adjusted accordingly. Therefore the sediment concentrations have been determined at these different flow conditions in order to use further in the modelling.

Table 7.13. Combined sediment concentration at the confluence under CBIO

Days	Total flow at confluence (m^3/s)	With minimum contribution of Q from PHLC (kg/m^3)	With maximum contribution of Q from PHLC (kg/ m^3)
30	17.0	0.180	0.200
60	19.4	0.178	0.200
90	21.7	0.199	0.200
120	23.1	0.214	0.390
150	23.1	0.324	0.870
180	23.1	0.366	0.880
210	24.6	0.363	0.880
240	23.1	0.316	0.870
270	23.1	0.233	0.870
300	17.0	0.097	0.480
330	17.0	0.086	0.480

7.10. Sediment transport downstream of RD 242 with combined sediment inflow from Machai Branch Canal and PHLC

The SIC Model at the moment does not accept sediment inflow from more than one point. It was therefore difficult to assess the combined effect of sediment inflow from both sources. Therefore, first the sediment discharge coming from PHLC was determined by simulations and then the combined sediment concentration was determined by adding the flow and sediment discharge from the Machai Branch Canal. Then the effect of different sediment transport scenarios has been analyzed the canals downstream of RD 242. On the basis of these sediment transport simulations, the sediment management options are discussed in next chapter. The combined sediment and flow data are given in Tables 7.11 and 7.12, which also has been used for sediment transport simulations for the canals downstream of RD 242.

7.10.1. At design discharge

Figure 7.13 presents modified bed levels (BL) downstream of RD 242 under design discharge with different sediment sizes. It is evident from Figure 7.13 that the fine sand tends to deposit in the upper reaches of the canal, whereas coarse silt tends to deposit in the head and middle reaches and medium silt deposits uniformly all over the canal, particularly in lower middle reaches. Deposition for fine sand, coarse silt and medium silt has been found to be 91,600 m^3, 43,200 m^3 and 21,000 m^3 respectively.

Figure 7.14 presents higher water levels due to raise in bed levels under fine sand and reduction in water discharge. As PI controllers respond to water levels, therefore as water levels keep on rising, discharge in the canal reduced correspondingly. This is one of the major operation and maintenance problems in automatic flow control systems. When sedimentation takes place at head reaches, the flow carrying capacity of the canal is reduced and it causes raise in the canal water level at flow control points. Then the water supplies to the canal start to reduce.

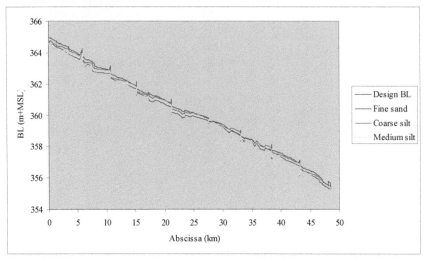

Figure 7.13. Sediment deposition downstream of RD 242 under design discharge

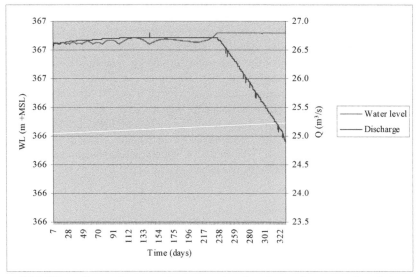

Figure 7.14. Water flow releases at the system headworks against water level changes in the canal

Figure 7.15 presents the AVIS gates response against the water level changes in the canal and flow releases at the system headworks. It can be seen that when the discharge at the canal headworks reduced to a certain level, some of the AVIS gates were opened fully and did not control flow anymore. This means, the required water levels, to regulate the AVIS gates, dropped down due to less supply from the source. Consequently, secondary offtakes started to draw less water than their design share. This phenomenon is difficult to manage in downstream controlled canals. As in case of local raise in water level, the offtake gate opening can be adjusted according to new water levels to maintain the design discharge. But once water levels start to drop and go down to certain extent, then the offtaking canals cannot draw their due share of water.

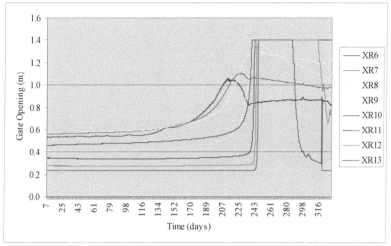

Figure 7.15. Gate openings against water flow releases at the system headworks and sedimentation in the canal.

7.10.2. At existing discharge

Some of the offtakes downstream of RD 242 were not operated at their design discharges due to low water requirements of the command area or poor maintenance conditions of the secondary canals. Therefore, these offtakes were supplied with less amount of water, whereas the flow was automatically adjusted at the system headworks in the main canal. This discharge was usually less than the design discharge; therefore sediment transport capacity of the canal was also reduced. Figure 7.16 presents variations in bed levels downstream of RD 242 under these flow conditions. The sediment transport under existing flow conditions and design conditions was almost same. Slightly more deposition took place under existing flow conditions, as the flows were reduced in July-August, which were peak sediment concentration periods. It can be seen from Figure 7.16 that there was more raise in bed levels in comparison with the bed levels raise under design discharge conditions. The total deposition under this scenario was about 48,900 m^3, 51,600 m^3, 29,300 m^3 for fine sand, coarse silt and medium silt respectively. In these simulations, the computations stopped after 270 days.

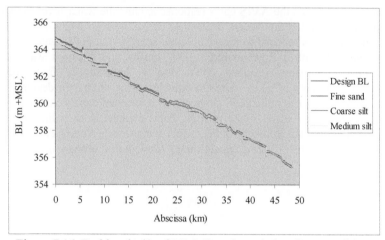

Figure 7.16. Bed levels d/s of RD 242 under existing flow conditions

7.10.3. At Crop Based Irrigation Operations

Variations in bed levels and under CBIO are given in Figures 7.17. It can be seen that fine sand deposited in the first two head reaches, whereas the other sediment sizes have not deposited much. Under CBIO, the flow releases started to reduce after about 250 days of simulation due to bed level raise and then subsequently, a water level raise was observed at the canal head. Once, the flow starts to reduce then it is comparatively easy to implement CBIO, as usually the flow is not required in its full quantity again in a one year period. Then during the canal closure the canal can be desilted and the capacity can be restored. But when AVIS gates stop to respond, then CBIO cannot be implemented anymore. The total deposition that took place under this scenario was 60,100 m³, 50,900 m³, 32,100 m³ for fine sand, coarse silt and medium silt respectively.

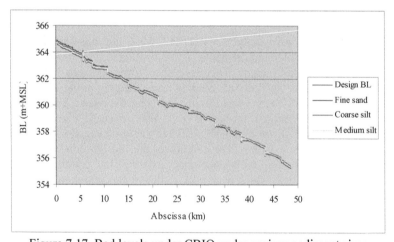

Figure 7.17. Bed levels under CBIO under various sediment sizes

8. Development of downstream control component in SETRIC Model

8.1. Background

A detailed discussion has been given on the use of sediment transport models in open channels in general and in irrigation canals in particular in Chapter 3. The one dimensional models like SOBEK (Deltares, 2009), MIKE 11 (Danish Hydraulic Institute, 2008), HECRAS (US Army Corps of Engineers, 2010), IALLUVIAL (Karim and Kennedy, 1982), Sediment and River Hydraulics-One Dimension (SRH-1D) Model (US Bureau of Reclamation, 2010), SEDICOUP (Belleudy and SOGREAH, 2000), FLUVIAL-12 (Howard, 1998; 2006) and HEC2SR (Simons et al., 1984) have been discussed there along with their applications.

It is evident from that discussion that almost all of the models have been developed for river flow simulation and morphological studies. Only a few of them can be applied to irrigation canals for this purpose. There are some commonly used models for steady and unsteady state flow simulations in irrigation canals like RootCanal (Utah State University, 2006), SIC (Baume, 2005) and CanalMan (Utah State University, 1997), but only the SIC Model, amongst these three models, is capable of modelling the sediment transport in irrigation canals.

Automatic flow controllers are considered as an integral part of models dealing with flow and sediment transport simulations in demand based systems. Generally most of the models do not have this automatic flow controller module. Therefore the application of such models on automatically controlled irrigation systems becomes difficult. Further, such simulations need fully dynamic models which can solve the Saint-Venant equations, which takes a lot of time during simulations.

Bringing all these factors together makes the job complicated and too difficult for the modellers and for the canal managers as well. Experience tells that most of the canal managers, particularly in least developed countries are not familiar with the above mentioned factors, particularly on canal automation. Therefore it is not easy for decision makers to use these models for decision support processes. Even if they become familiar with real time flow simulation models coupled with unsteady state flow simulations, it will not be a simple task to simulate the sediment transport in irrigation canals under study due to complicacies embedded in combining all these factors. Therefore it is useful to develop a simplified model which would be able to simulate flow and sediment transport in supply and demand based irrigation canals with fixed and automatic flow operations.

To cope with these problems of flow and sediment transport simulations in demand based irrigation canals, the mathematical model SEdiment TRansport in Irrigation Canals (SETRIC) has been developed. The conceptual background of this model was given by Mendez (1998) with the focus of sediment transport modelling in irrigation canals. Later, Paudel (2002; 2010) developed and tested the model for supply based irrigation canals. He successfully applied the model for sediment transport studies in the

Sunsari Morang Irrigation Scheme in Nepal. In this study the SETRIC Model has been improved with the incorporation of the downstream control system and applied to the Lower Machai Branch Canal and Maira Branch Canal Pakistan.

8.2.   Rationale

Hydraulics of the downstream control systems are somehow different from the traditional upstream controlled irrigation systems. The hydraulic models developed for upstream control generally cannot be applied for downstream control systems. As in upstream control, with the reduction in flow the water levels in the canal also drop accordingly. But in downstream control systems the water levels must increase with a decrease in flow. The upstream control systems are usually manually operated, while the downstream control systems are mainly automatically controlled. In upstream control systems the target water level is set at upstream of the flow and water level regulating structures, while in downstream control systems these are fixed at the downstream of the regulating structures. Therefore the flow control algorithms developed for upstream control cannot be directly applied to downstream control.

Further, downstream control systems are generally applied to demand based irrigation. The hydrograph at the canal headworks remains varying in response to the operation of the secondary or tertiary irrigation offtakes. In steady state models, by reducing the flow at the headworks causes a drop in the water levels in the irrigation canals and vice versa, which is in principle contradictory to the flow control methods in downstream control systems. In downstream control systems any decrease in flow in the canal causes an increase in water levels in the canal and vice versa. The increase in water level provides water storage in the canal, which is utilized for diminishing the effect of canal filling times or response times when an offtake is opened again after closure. This storage provides immediate water supply to the secondary or tertiary system without decreasing flows to the other offtaking channels, which may otherwise be faced in case of some upstream manually fixed control mechanism.

In the SETRIC Model this problem has been solved and there is no need to give the flow hydrograph at the canal headworks. Only design discharge is needed at the canal headworks. Then according to the operation of the secondary or tertiary offtakes the flow hydrographs are automatically adjusted. The water levels in the canal then respond to flow variations in the system and increase or decrease accordingly. As the hydrograph is generated automatically, it eliminates the need of flow control modules at the system headworks for steady state modelling. Incorporation of these options make the SETRIC Model a simple and easy tool to apply for flow and sediment transport modelling in upstream and downstream controlled irrigation canals for steady state modelling.

The main objective of this exercise is to describe the functioning of automatically downstream controlled irrigation canals. Then to address the above mentioned complications in hydraulic and sediment transport modelling in such canals by developing a simplified steady state model. Then applying the developed model to the irrigation canal under study.

8.3.   SETRIC Model

SETRIC is a one-dimensional mathematical model which simulates flows and sediment transport. The flows are simulated as quasi-steady and gradually varied, where sediment transport is simulated in equilibrium as well non-equilibrium conditions. Further description of the model has been given as under for general hydraulic and sediment transport computations where the details on downstream control system are given in section 3.8.

8.3.1. Water flow calculations

Water flow calculations in the model are based on sub-critical, quasi-steady, uniform or non-uniform conditions (gradually varied flow). The predictor-corrector method is used to calculate the flow profile. The length step selected is taken constant throughout the canal length, but if the length of the section is less than the length step or a multiple of the length step then the step is modified accordingly. Accuracy of the prediction of water depth used in the model is 0.005 m.

For the quasi-steady and uniform flow conditions continuity equation becomes

$$\frac{\partial Q}{\partial x} + q_i = 0 \qquad (8.1)$$

Similarly the dynamic equation becomes:

$$\frac{dy}{dx} = \frac{S_o - S_f}{1 - Fr^2} \qquad (8.2)$$

Where
Q	= flow rate (m³/s)
q_i	= later inflow/outflow discharge (m³/s)
y	= water depth (m)
So	= bottom slope (m/m)
S_f	= Slope of energy line (m/m)
Fr	= Froude number
x	= length co-ordinate in x-direction (m)

The dynamic equation of the flow profile has been solved numerically by using the predictor-corrector method.

Predictor-Corrector method

Mendez (1998) describes the solution of Equation (8.2) as follows:

Computation starts from downstream end point ($x=x_o$) by computing the derivative of Equation 8.2 at point $x = x_i = x_o$ for given S_o, S_f and Fr.

Figure 8.1 presents a schematization of the predictor-corrector method for the numerical solution of the GVF equation.

Water depth, y_i, at point $x = x_{i+1}$ is calculated as:

$$y_{i+1} = y_i + (\frac{dy}{dx})_i (x_{i+1} - x_i)$$ (8.3)

Then S_{fi+1} and Fr_{i+1} are calculated and the derivative at point $x = x_{i+1}$ is taken as:

$$(\frac{dy}{dx})_{i+1} = \frac{(S_o - S_f)_{i+1}}{(1 - Fr^2)_{i+1}}$$ (8.4)

Then mean derivative is calculated as:

$$(\frac{dy}{dx})_{mean} = \frac{(\frac{dy}{dx})_i + (\frac{dy}{dx})_{i+1}}{2}$$ (8.5)

and new depth y_{i+1} is calculated by:

$$y_{(i+1)2} = y_i + (\frac{dy}{dx})_{mean} (x_{i+1} - x_i)$$ (8.6)

and accuracy of the predictor-corrector method is checked by

$$\left| y_{(i+1)1} - y_{(i+1)2} \right| \le e$$ (8.7)

Where e = degree of accuracy.

8.3.2. Roughness calculations

In alluvial channels, the boundary friction is related to the skin friction (grain related) and to the form losses caused by the bed forms. Effect of bed forms is particularly substantial with ripples and dunes. In channels with sediment laden flows the determination of roughness becomes complicated when bed forms are developed in the channel beds. These bed forms affect channel's hydraulic resistance significantly. Roughness conditions in the bed are simulated by using an equivalent height of the sand roughness, k_{se}, which is equal to the roughness of the sand that gives the same resistance as the bed form. Hence the composite height of roughness becomes equal to the summation of the roughness due to grain and the roughness due to bed forms developed in the channel bed.

$$k_{se} = k_s' + k_s''$$ (8.8)

Where
k_{se} = Equivalent roughness (m)
k_s' = roughness due to grain (m)
k_s'' = roughness due to form developed in the bed (m)

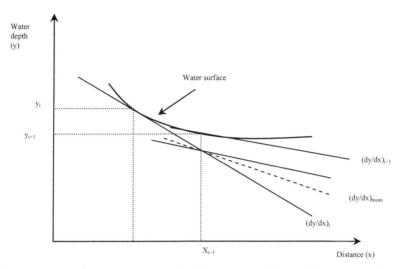

Figure 8.1. Predictor-corrector method for water surface profile computations

In SETRIC the Van Rijn method is used to determine the equivalent roughness in the bed. Van Rijn (1993) divided the hydraulic flow regime, based on the roughness conditions, into three stages: smooth, rough and transition. As the total equivalent roughness is due to grain size of the bed material and bed form created due to particle movement. The equivalent roughness is determined for two conditions: without the movement of particles and with the movement of particles.

8.3.3. Determination of roughness on the side walls

The model takes into account the roughness of the sidewalls. The sidewall's roughness is mainly due to vegetation because sediments seldom settle on the side slopes as the side slopes angle is usually larger than the angle of repose of the material. Hence it can be safely assumed that there is no settlement in the sides (Paudel, 2002). Then, the roughness in the sides becomes due to type of material and the vegetation, that may be present at the beginning or likely to grow during the period of canal operations, depending upon the type of maintenance.

The problem has been divided into two categories: (i) sides without any vegetation and (ii) the other is side covered with vegetation. The second case has been further divided into various weed factors from 0.1 (densely grown) to 1.0 (ideally clean) as formulated by Mendez (1998). For ideally maintained conditions a weed factor from 0-5% can be used, its value may vary from 0.9 to 1.0. For fully-grown vegetation the 75% weed factor can be used, it ranges between 0.1-0.3.

Maintenance conditions also have been considered and are divided into three possible maintenance scenarios: no maintenance, well maintenance and ideal maintenance. For poorly maintained canals, at each time step starting from the beginning the effect of vegetation will be accounted in the roughness calculation. It becomes maximum if the simulation period equals or exceeds the total growing period of vegetation. For well maintained canals, the effect of vegetation on the roughness is considered from the beginning. Weed factor from 0-5% is taken for the initial stage and it increases linearly up to the periodic maintenance time. Mendez (1998) defined the periodic maintenance interval at two months approximately, however it depends upon the type of vegetation in terms of total growing time and its effect on roughness.

8.3.4. Galappatti's depth integrated model

In SETRIC the diffusion model is used. Galappatti (1983) developed a depth integrated model to solve the 3-D convection-diffusion equation with following assumptions:

- diffusion terms other than vertical are neglected:
- concentration is expressed as the depth averaged concentration.

With the above assumptions the three dimensional convection-diffusion equation reduces to:

$$\frac{\partial c}{\partial t} + u\frac{\partial c}{\partial x} + w\frac{\partial c}{\partial z} = w_s\frac{\partial c}{\partial z} + \frac{\partial}{\partial z}\left(\epsilon_z\frac{\partial c}{\partial z}\right) \tag{8.9}$$

For steady uniform flow where the suspended sediment is not in equilibrium, the variation in concentration can be written as:

$$\overline{c_e} = \overline{c} + T_A\frac{\partial \overline{c}}{\partial t} + L_A\frac{\partial \overline{c}}{\partial x} \tag{8.10}$$

This equation is the first order solution of the convection diffusion equation of 2D. If the sediment flow is steady the above equation can be written as:

$$\overline{c_e} = \overline{c} + L_A\frac{\partial \overline{c}}{\partial x} \tag{8.11}$$

Integrating the above equation for x from 0 to x and c from c to c_o to c gives:

$$\int_0^x dx = L_A\int_{c_o}^{\overline{c}}\frac{d\overline{c}}{\overline{c_e} - \overline{c}} \tag{8.12}$$

$$\overline{c} = \overline{c_e} - (\overline{c_e} - \overline{c_o})*e^{-\frac{x}{L_A}} \tag{8.13}$$

Where

$\overline{c}$ = required concentration at a distance x from the origin

$\overline{c_e}$ = equilibrium concentration at x from the origin

$\overline{c_o}$ actual concentration at the origin

L_A = adaptation length (m)
x = distance from the origin (m)
f = coefficient (values given by Galappatti, 1983)

Galappatti's depth integrated model is for suspended load only. Assuming that the adaptation characteristics for suspended sediment to be comparable with the adaptation characteristics of bed movement, the same model can be proposed for total load transport. Equation 8.13 can be written in the form of total sediment concentration as:

$$c = c_e - (c_e - c_o)\exp^{-\frac{x}{L_A}} \qquad\qquad (8.14)$$

with:

$$L_A = f\left(\frac{u_*}{U}, \frac{w_s}{u_*}, h\right)$$

Where

c = total sediment concentration at distance x (ppm)
c_e = total sediment concentration in equilibrium conditions (ppm)
c_o = total sediment concentration at origin (ppm)
h = water depth (m)
L_A = adaptation length (m)
u_* = shear velocity (m/s)
U = mean flow velocity (m/s)
w_s = fall velocity (m/s)

8.3.5. Separation of bed and suspended load

The model SETRIC does not simulate the bed load and suspended load separately but it simulates the total sediment load. If the structure is with raised crest then it is possible that the bed load part of the total sediment is trapped upstream of the structure. Then it means only the suspended load will move downstream of the structure. Out of the total load the suspended load carrying capacity of the canal is calculated using the following relation (Mendez, 1998):

$$q_{s,s} = \frac{q_{s,t}}{\left[\frac{5}{12} * \left(\frac{d_{50}}{y}\right)^{0.2} * D_*^{0.6} + 1\right]} \qquad\qquad (8.15)$$

Where

$q_{s,s}$ = suspended part of the sediment transport rate (m²/s)
$q_{s,t}$ = total load transport rate predicted (m²/s)
d_{50} = mean sediment size (m)

$$y \qquad = \text{water depth (m)}$$
$$D_* \qquad = \text{dimensionless grain parameter}$$

8.3.6. Concentration downstream of inflow and outflow points

In case of lateral inflow or outflow in the system the downstream concentration level will change accordingly depending upon the concentration and discharge in the inlet or outlet. It is given by:

$$c_2 = c_1 \frac{(Q_2 - Q_i)}{Q_2} + c * \frac{Q_i}{Q_2} \tag{8.16}$$

Where

c_1 = sediment concentration in the canal before the inflow (ppm)
c_2 = sediment concentration after inlet (ppm)
c_i = concentration of inflow (ppm), at upstream boundary
Q_i = inflow discharge (m³/s)
Q_2 = total discharge after the inlet (m³/s)

In case of outflow from the system, the sediment concentration after offtake point is given by:

$$c_2 = \frac{Q_1}{(Q_2 - Q_o * f_d)} * c_1 \tag{8.17}$$

Where

Q_1 = discharge in the canal before the off-take (m³/s)
Q_o = off-take discharge (m³/s)
f_d = distribution ratio of sediment within lateral and main

8.3.7. Morphological changes in the bed

Following equations give the relationship between water movement and the morphological changes in the bed level:

$$\frac{dh}{dx} = \frac{S_o - S_f}{1 - Fr^2} \tag{8.18}$$

and:

$$\frac{\partial Q_s}{\partial x} + B(1 - p_r)\frac{\partial z_b}{\partial t} = 0 \tag{8.19}$$

These equations are solved alternatively. First the water flow equations are solved (using the predictor-corrector numerical method) and then the results of the calculations are used in the next step to solve the sediment transport equation. The modified lax method has been used to solve the second equation.

Modified Lax Method

The sediment continuity equation can be solved by using implicit or explicit numerical schemes. Lax, modified lax, leap frog, lax-Wendorff are some of the explicit finite difference schemes. In this study the Modified Lax Method has been used, which is given as:

$$z_{i,j+1} = z_i + \left[\frac{1}{B(1-p_r)} \frac{(Qs_{i+1,j} - Qs_{i-1,j})}{2\Delta x} - \frac{1}{4\Delta t} \left\{ \begin{array}{l} (\alpha_{i+1,j} + \alpha_{i,j})(z_{i+1,j} + z_{i,j}) \\ -(\alpha_{i,j} + \alpha_{i-1,j})(z_{i,j} - z_{i-1,j}) \end{array} \right\} \right] \Delta t \quad (8.20)$$

This numerical scheme cannot applied to the downstream and upstream boundaries. An adapted scheme to the downstream boundary is described by:

$$z_{i,j+1} = z_i - \left[\frac{1}{B(1-p_r)} \frac{(Qs_{i,j} - Qs_{i-1,j})}{\Delta x} + \frac{1}{2\Delta t} \left\{ (\alpha_{i+1,j} + \alpha_{i,j})(z_{i,j} - z_{i-1,j}) \right\} \right] \Delta t \quad (8.21)$$

and the upstream boundary condition is described by:

$$z_{i,j+1} = z_i - \left[\frac{1}{B(1-p_r)} \frac{(Qs_{i+1,j} - Qs_{i,j})}{\Delta x} - \frac{1}{2\Delta t} \left\{ (\alpha_{i+1,j} + \alpha_{i,j})(z_{i+1,j} - z_{i,j}) \right\} \right] \Delta t \quad (8.22)$$

Where

i	= subscript for space
j	= subscript for time
B	= Bed width (m)
p_r	= Porosity
Qs	= sediment discharge (m^3/s)
z_b	= bed level (m)
α	= parameter used for stability and accuracy of the numerical scheme
Δt	= time step
Δx	= length step (m)

The stability of the scheme is given by (Vreugdenhil, 1989):

$$\sigma^2 \le \alpha \le 1 \quad (8.23)$$

Equation 8.21 is the general equation for an explicit scheme and by giving different values to α different schemes can be made. A scheme of the intermediate type can be found if (Vreugdenhil and DeVries, 1967):

$$\alpha = \sigma^2 + \beta \tag{8.24}$$

The accuracy of the modified Lax Scheme can be adjusted by means of the parameter β (Abbot and Cunge, 1982). Accuracy of this scheme is increased if $\beta = 0.01$ (Vreugdenhil and Wijbenga, 1982):

$$\alpha \approx \sigma^2 + 0.01 \tag{8.25}$$

σ is called the Courant number and is described as:

$$\sigma = NV \frac{Qs/Q}{1 - Fr^2} \frac{\Delta t}{\Delta x} \tag{8.26}$$

Where
Fr	= Froude number
N	= exponent of velocity in the sediment transport equation
V	= mean flow velocity (m/s)
Q	= water flow rate (m^3/s)
Q_s	= sediment transport rate (m^3/s)
Δt	= time interval
Δx	= length step (m)

8.3.8. Boundary conditions

The boundary condition for water flow calculations is the head discharge, upstream boundary condition and the downstream boundary condition is defined by the water level. The tail water level is not changed during one simulation period. Similarly for sediment transport calculations the inflow sediment concentration is the upstream boundary. It is not possible to change the inflow sediment concentration level throughout the simulation period. There is no sediment deposition at the structures, means the sediment inflow rate into a structure will be equal to the sediment outflow rate.

8.4.   Improvements in the SETRIC Model

So far the SETRIC Model is capable of sediment transport simulations only in upstream controlled irrigation canals. As mentioned earlier that the models developed for hydraulic and sediment transport simulations for upstream controlled irrigation canals cannot be applied to downstream controlled demand based irrigation canals due to different types of flow control mechanism in upstream controlled and downstream controlled irrigation canals.

To deal with these limitations the downstream control component has been added in the SETRIC Model, which makes it a quite simple and straightforward tool for hydraulic and sediment transport simulations in demand based downstream controlled irrigation canals. A detailed description on downstream controlled irrigation canals has been given in Chapter 3, whereas the description and working mechanism of downstream control gates will be elaborated in the following sections in order to know their function in the model.

8.4.1. AVIS and AVIO gates

Downstream control canals are equipped with AVIO and AVIS gates and are used to control water levels at the canal headworks and in canal reaches. The gates maintain a constant water level at their downstream side (upstream end of a canal pool). These are hydro-mechanical self-operating gates. The AVIS and AVIO gates are similar gates. The name 'AVIS' has a French background: *AV* is from "*aval*", means downstream, and S is from "*surface*", whereas in "AVIO" the letter "*O*" is from "*Orifice*". It illustrates that the AVIS gate operates at a free surface flow and AVIO gate operates under orifice conditions as shown in Figure 8.2. Further, two types of AVIS/AVIO gates are available: the "High Head" type and the "Low Head" type. The High Head gates have a narrower gate than the Low Head gates with the same float. The high head gates are usually employed in the irrigation canals where narrow canal cross-sections are required. The choice between the open type (AVIS gates) and the orifice-type (AVIO gates) is solely determined by the maximum headloss likely to occur between the upstream and downstream-controlled water levels.

A. A pair of AVIS gates   B. A pair of AVIO gates
Figure 8.2. AVIS and AVIO gates in action

8.4.2. Gate index of AVIS/AVIO gates

The *AVIS* and *AVIO* gates are indentified by their dimension indices "*r/b*" and "*r/a*", respectively. The gate index *r* is the float radius in cm, whereas the index *b* of the AVIS gates equals the sill width of the structure in cm and the index *a* of the AVIO gates equals the area of the orifice of the structure in cm^2. High head and low head AVIS/AVIO gates are available in a series of increasing dimensions. Ankum (undated) uses the gate index "*r*" for identification of the AVIS/AVIO gates, as given in Table 8.1.

Table 8.1. The available gate types of the AVIS/AVIO gates (Ankum, undated)

Gate index "r"	AVIS "r/b"		AVIO "r/a"	
	High Head	Low Head	High Head	Low Head
28			28/6	
36			36/10	
45			45/16	45/32
56	56/106		56/25	56/50
71	71/132		71/40	71/80
90	90/170	90/190	90/63	90/125
110	110/212	110/236	110/100	110/200
140	140/265	140/300	140/160	140/315
160	160/300	160/335	160/200	160/400
180	180/335	180/375	180/250	180/500
200	200/375	200/425	200/315	200/630
220	220/425	220/475	220/400	220/800
250	250/475	250/530	250/500	250/1000
280	280/530	280/600	280/630	280/1250

AVIS/AVIO gates are water level regulators and not discharge regulators. So, they are not water-tight. If operating conditions occasionally require the flow to be shut off completely, the AVIS gate needs to be provided with stoplogs. Figure 8.3 presents a structural set up for an AVIS gate. Whereas the AVIO gate is provided with an upstream orifice with emergency gate. .

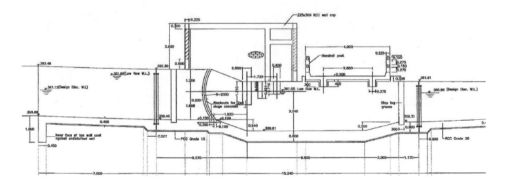

Figure 8.3. Structure of an AVIS gate (180/335)

8.4.3. Hydraulics of AVIS/AVIO gates

The manufacturer of the AVIS/AVIO gates provides headloss charts, Figure 8.4, for the selection of these gates (Ankum, undated). These charts show the relationship between discharge Q and the head J of the upstream water level above the gate axis.

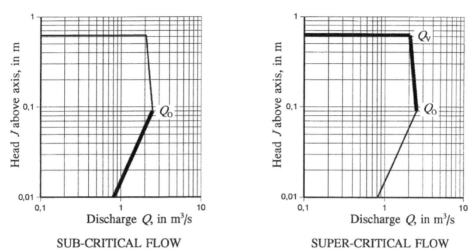

SUB-CRITICAL FLOW   SUPER-CRITICAL FLOW
Figure 8.4. Headloss charts for AVIS/AVIO gates (Source: Ankum, undated)

8.4.4. Discharge computation

The discharge computation through the AVIS/AVIO gates is a two step procedure. First of all the gate opening is calculated based on the downstream water level and then the discharge is computed on the basis of the gate opening.

The gate opening is calculated using the downstream water level and the design and tuning parameters.

Difference between axis elevation h_a and the downstream water elevation h_2 is given by:

$$\varepsilon_{act} = h_a - h_d \tag{8.27}$$

Maximum angle of gate opening, α_{max} is given by:

$$\alpha_{max} = a\sin\left(\frac{h_a}{R}\right) - a\sin\left(\frac{h_a - h}{R}\right) \tag{8.28}$$

Where
ε_{act}	= actual decrement (m)
h_d	= downstream water elevation (m)
R	= gate radius (m)
h_a	= axis elevation above sill (m)
α_{max}	= gate opening angle (radians)
h	= maximum vertical gate opening

Actual gate opening angle α is calculated as if:

$$
\begin{cases}
(\varepsilon_{act} > \varepsilon_{des}): \alpha = \alpha_{max} \\
(\varepsilon_{act} < \varepsilon_{des}): \alpha = 0 \\
(\varepsilon_{act} \leq \varepsilon_{des})\,and\,(\varepsilon_{act} \geq 0): \alpha = a\sin\left(\dfrac{a}{d}\sin(\alpha_{max})\right)
\end{cases}
\tag{8.29}
$$

Which means that the gate is fully opened in case of $\varepsilon_{act} > \varepsilon_{des}$ or that the gate is almost fully closed for $\varepsilon_{act} = 0$ and in case of $\varepsilon_{act} > 0$ and $\varepsilon_{act} < \varepsilon_{des}$, the gate opening depends upon the difference between ε_{act} and ε_{des}.

From the α value, the relative vertical gate opening w can be calculated as:

$$
w = \left\{\left(R^2 - h_a^2\right)^{0.5}\sin(\alpha) + h_a(1 - \cos(\alpha))\right\}/w_{max}
\tag{8.30}
$$

After calculating the gate opening the discharge through the gate is calculated as:

$$
Q = 4.1wR^2\sqrt{J}
\tag{8.31}
$$

Where

Q	= discharge (m^3/s)
w	= vertical gate opening (m)
w_{max}	= maximum vertical gate opening (m)
R	= gate radius (m)
J	= head of the upstream water level above the gate axis (m)
h_a	= axis elevation above sill (m)
α_{max}	= gate opening angle (radians).

This discharge equation can be used to calculate the discharge through the high head and low head AVIS gates, AVIO gates and the AMIL gates (Baume, 2005).

The relationships for AVIS high head gates installed in the irrigation canals under study are given by in relation to gate radius R as:

- $J_{max} = 0.45R$
- Axis elevation above sill = $0.64R$
- Maximum vertical gate opening = $0.56R$
- Decrement = $0.03R$

8.5.   Computation procedure in SETRIC Model for downstream control

The computation procedure for the SETRIC model for downstream control is as follows:

- Computations start from the downstream end at known flow and water level conditions;
- Then the water surface profile is computed from the tail to the last (first from the tail) cross regulator taking into account the other factors like offtake withdrawal etc. The differential equation of water surface profile is solved by using Predictor-Corrector Method (See previous chapters);
- Then on the basis of downstream water level the actual decrement ε_{act} is calculated, which is the difference between axis elevation h_a and the downstream water elevation, h_d;
- On the basis of the actual decrement the gate opening angle α is calculated;
- Then on the basis of gate opening angle α the actual gate opening w is calculated;
- Then on the basis of the actual gate opening α and discharge in the canal the upstream water level is computed;
- Upstream water level is then taken as the boundary condition for next canal reach;
- This procedure continues till the upstream end of the canal (the first reach);
- Then sediment transport computations start from the upstream end and go to downstream end of the canal.

8.6. Application of the SETRIC model to automatically downstream controlled irrigation canal

SETRIC model was applied to the automatically downstream controlled irrigation canal in Pakistan in order to study the sediment transport in these irrigation canals and to verify the model's application to downstream control. The irrigation system downstream of the RD 242 comprises of Lower Machai Branch Canal and the Maira Branch Canal. Main features of these irrigation canals under study have been given in Table 5.1.

The application of the SETRIC Model to upstream controlled irrigation canals has been evaluated by the Paudel (2010) by comparing the model results with the existing widely used flow and sediment transport models. For hydraulic modelling the SETRIC model was compared with DUFLOW Model (IHE, 1998) and for sediment transport simulations comparison the model was compared with SOBEK-RIVER Model (Delft Hydraulics and Ministry of Transport, Public Works and Water Management, 1994b). He found quite good proximity between the predicted values of the SETRIC Model with the DUFLOW and the SOBEK-RIVER Models.

In this study the model was applied for four different flow conditions, the design flow, the existing flow, the 50% of design flow and canal operations under CBIO for hydraulic and sediment transport simulations. The model was calibrated according to the field measurements in terms of canal roughness. The canal roughness was changed to meet the measured and simulated water levels and for sediment transport calibration, a multiplication factor with the equilibrium sediment transport formula was used.

8.6.1. Calibration and validation of the model

All of the inflow and outflow from the canal was measured for calibration purposes. The canal was kept at steady state conditions during the measurements. The discharges in the canal were measured with the current meter and the discharges of the offtakes were determined by reading the water levels having already calibrated the gauges. The canal was divided into three parts for discharge measurement purposes. The discharge was measured at an abscissa of 500 m from the confluence, then at 16,500 m from the confluence and then at an abscissa of 43,200 m from the confluence point. Then the water levels at upstream and downstream of every cross regulator were measured. The Manning's roughness value in the model was changed in order to meet the measured and simulated values. There was a nominal difference in the design roughness value and the calibrated roughness values. It might be due to the newly built canal or regular maintenance of the canal. The results of model calibration and validation are given in Table 8.2. These values of calibration and validation parameters suggest that the model is suitable for application.

Table 8.2. Flow calibration and validation values for SETRIC Model

S. No	Parameter description	Calibration	Validation
1	ME (m)	0.182	0.196
2	RMSE	0.103	0.112
3	EF	0.997	0.994
4	CRM	-0.001	-0.003
5	MAE (%)	-0.103	-0.117

8.7.   Flow simulations

8.7.1. Flow simulation under design flow conditions

The total design discharge at the confluence point is 31.9 m^3/s but the canal usually runs at 25.5 m^3/s. There are two proposed lift irrigation schemes in the Maira Branch canal, which would draw a total 3.4 m^3/s discharge. Water has been allocated to these lift schemes but they have not been built yet. There are two other secondary canals, the Pehure Branch Canal, with design discharge of 2.95 m^3/s and the Link Channel with design discharge of 3.06 m^3/s. These two secondary canals were not remodelled during the remodelling of this irrigation system in 2002, so they cannot carry their design discharge. Their old design capacities are half of their new design discharges. Therefore each of these offtakes can carry a maximum of 1.5 m^3/s discharge. So in total they carry about 3.0 m^3/s discharge. Under these circumstances, the design discharge or maximum discharge at the confluence point becomes about 25.5 m^3/s. So in these simulations this discharge has been considered as design discharge.

Figure 8.5 presents water surface elevations along the canal at design discharge. Simulation results show that the model has accurately simulated the flows at design discharges along the canal. The water levels upstream and downstream of the cross regulators are relatively higher than the design water levels at low flows. It is because the discharge at the canal head is less than the actual design discharge, which is 31.9 m^3/s. The resulting water levels at cross regulators, head above the axis and gate openings are given in Table 8.3.

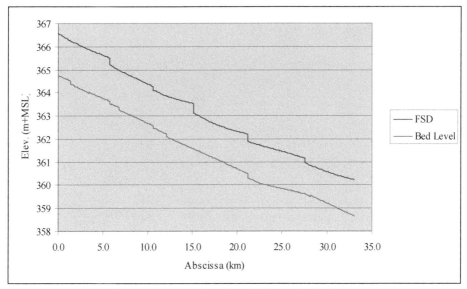

Figure 8.5. Water levels at full supply discharge

Table 8.3. Water levels and gate opening at full supply discharge

Cross regulator	Gate type	Abscissa	Water level	Q	Axis level	Head above axis	Gate opening
		(m)	(m+MSL)	(m^3/s)	(m+MSL)	(m)	(m)
XR-6	200/375	5,770	365.51	23.1	365.30	0.21	1.25
XR-7	180/335	10,680	364.25	20.1	364.19	0.16	1.56
XR-8	180/335	15,200	363.53	18.7	363.19	0.34	1.00
XR-9	180/335	21,170	362.21	15.0	361.98	0.23	0.97
XR-10	160/300	27,570	361.14	13.2	361.05	0.09	1.70
XR-11	160/300	33,070	360.24	10.3	359.96	0.28	0.76

8.7.2. Flow simulation at 50% of full supply discharge

The other main scenario in this irrigation system is the 50% flow of the full supply discharge (FSD) or design discharge. As already mentioned, the flows in the main canal are not allowed to go lower than the 50% of the FSD. Therefore simulations were performed by closing half of the secondary offtakes whereas the direct outlets remained open. The refused flow becomes about 12.5 m^3/s. Figure 8.6 presents water surface elevations along the canal at this flow condition. The simulation results of 50% of FSD seem quite close to expected water levels, which should be somewhere in between the water levels at full supply discharge and water levels at zero flow. The gate openings, head above the axis and water levels on cross regulators under these flow conditions are given in Table 8.4.

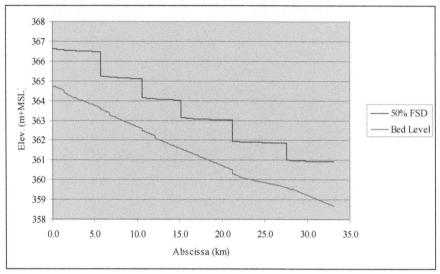

Figure 8.6. Water levels along the along under 50% flow of FSD

Table 8.4. Water levels and gate openings at 50% of full supply discharge

X-regulator	Gate type	Abscissa	Water level	Discharge	Axis level	Head above axis	Gate opening
		(m)	(m+MSL)	(m³/s)	(m+MSL)	(m)	(m)
XR-6	200/375	5,770	366.46	13.5	365.30	1.16	0.31
XR-7	180/335	10,680	365.10	11.2	364.19	0.91	0.36
XR-8	180/335	15,200	364.03	10.6	363.19	0.84	0.36
XR-9	180/335	21,170	363.02	7.1	361.98	1.04	0.22
XR-10	160/300	27,570	361.85	6.3	361.05	0.80	0.27
XR-11	160/300	33,070	360.92	5.2	359.96	0.96	0.21

8.7.3. Flow simulation at 75% of full supply discharge

As in CBIO the canal is operated at many flow conditions. Therefore another scenario was simulated by keeping the flow at 75% of FSD. Figure 8.7 presents water levels against the three simulated scenarios, which are 100%, 75% and 50% discharges of FSD. Water levels on 100% flow, 50% flow are given in Tables 8.3 and 8.4, whereas water levels and corresponding gate openings for 75% flow are given in Table 8.5.

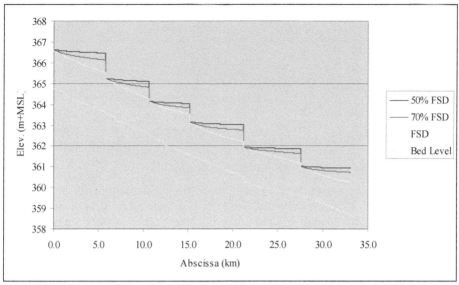

Figure 8.7. Water surface profiles at different discharges

Table 8.5. Water levels at cross regulators at a discharge of 75% of full supply discharge

X-regulator	Gate type	Abscissa	Water level	Q	Axis level	Head above axis	Gate opening
		(m)	(m+MSL)	(m³/s)	(m+MSL)	(m)	(m)
XR-6	200/375	5,770	366.12	19.5	365.30	0.82	0.53
XR-7	180/335	10,680	364.80	16.5	364.19	0.61	0.66
XR-8	180/335	15,200	363.82	15.3	363.19	0.63	0.60
XR-9	180/335	21,170	362.76	11.9	361.98	0.78	0.42
XR-10	160/300	27,570	361.62	10.0	361.05	0.57	0.52
XR-11	160/300	33,070	360.72	7.8	359.96	0.76	0.35

8.8. Sediment transport simulations

8.8.1. Calibration and validation of SETRIC Model for sediment transport

The SETRIC Model was calibrated for sediment transport by using existing sediment inflow to the canal. A multiplication factor was used with the sediment predictor for model calibration. The simulation results were measured with the cross sectional survey conducted in January 2008. Then the model was validated by using sediment inflow data from March 2008 to July 2008. Then the bed evolution was compared with the cross section survey conducted in July 2008. The results of the SETRIC Model calibration and validation are presented in the Table 8.6.

Table 8.6. Flow calibration and validation values for SETRIC Model

S. No	Parameter description	Calibration	Validation
1	ME (m)	0.373	0.419
2	RMSE	0.133	0.160
3	EF	0.993	0.981
4	CRM	-0.004	-0.005
5	MAE (%)	-0.114	-0.121

8.8.2. Sediment inflow to the irrigation canal under study

The sediment concentration entering into lower Machai Branch Canal at RD 242 is given in Table 7.5. Silt is the dominant factor, ranging from 60 to 70 % in the sediments characteristics, sand is the other major portion, whereas clay has small portion ranging between 4 to 10 % of the total load.

Along with hydraulic computations the SETRIC Model is equally capable of sediment transport simulations. The flow and sediment transport computations are performed at every time step. In one step the hydraulic computations are performed and then the sediment transport computations are done according to the hydraulic conditions in that time step. So the effects of flow changes can be seen on the sediment transport behaviour and also the effect of sediment transport can be observed on changes in canal hydraulics. Sediment transport in the SETRIC Model has also been simulated under the already mentioned three major canal operation scenarios i.e. the design discharge conditions, existing flow conditions and the CBIO. The simulations have been performed with the existing sediment inflow rates and characteristics, as given in Table 7.5.

8.8.3. Under design discharge with existing sediment inflow

Figure 8.8 presents sediment transport under design flow. The sediment inflow data has been used which is given in Table 7.5. A total deposition has been resulted as 14,000 m^3 which is quite close to the sediment deposition in selected reaches of the canal simulated by the SIC Model, given in Chapter 7.

8.8.4. Under existing water and sediment inflow

Figure 8.9 presents sediment transport under existing flow conditions with existing sediment inflow. The sediment data as given in Table 7.5 has been used for these simulations. A total deposition of 18,600 m^3 has been observed, which is more than the sediment deposition under design conditions. The predicted sediment deposition is fairly close to the measured sediment deposition. These results further strengthen the model's calibration and validation by showing a close match between the simulated values of sediment deposition and the measured values.

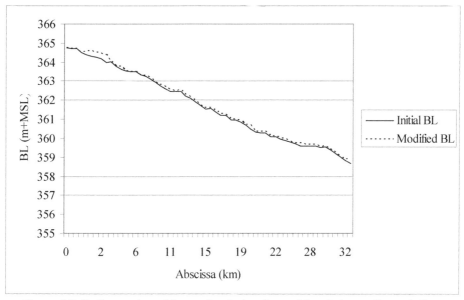

Figure 8.8. Sediment deposition under design flow with existing sediment inflow

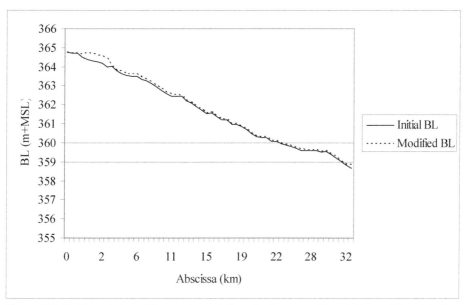

Figure 8.9. Sediment deposition under existing discharges with existing sediment inflow

8.8.5. Under CBIO and existing sediment inflow

The results of sediment transport under CBIO with existing sediment inflow are given in Figure 8.10. The sediment deposition under CBIO is highest, more than the sediment deposition under canal operation with design discharges and with existing flow condition. A total deposition volume of 23,100 m^3 has been observed, which is mainly concentrated in the upper canal reaches.

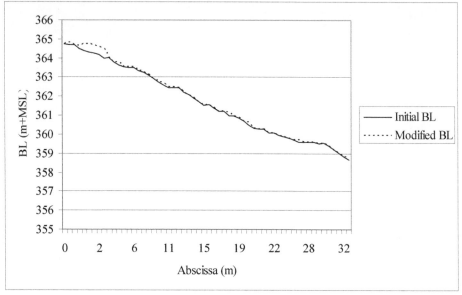

Figure 8.10. Sediment deposition under CBIO with existing sediment inflow conditions

9. Managing Sediment transport

9.1. Traditional approach of sediment management in Indus Basin Irrigation System

In IBIS sediment management has been in practice since the inception of the irrigation system. Almost all of the effective techniques are applied for this purpose including design and operation of the irrigation canals and regular desilting of the canals on annual basis. Every year in the month of January the whole irrigation system is closed and almost all of the silt affected secondary irrigation canals are desilted, depending upon the availability of funds. In addition to desiltation, other canal maintenance works are also executed like berm strengthening, vegetation removal, structures repair, etc.

From canal design point of view, generally all of the irrigation canals are designed on Lacey's regime concept in IBIS. In this concept of canal design the flow velocities in the canals are maintained in such a way that they neither scour nor deposit sediment in the canal prism during canal operation. In the beginning some scouring and deposition take place in the canal but with the passage of time hydraulic characteristics of the canal are adjusted in such a way that theoretically neither of the phenomena takes place. Later many researcher in Pakistan and abroad conducted research on Lacey regime concept improved it by considering many additional factors. For example Blench (1957) gave two different silts factors for canal bed and side slope, whereas Lacey (1929) gave only average silt factor for canal bed and side slope. Lacey did not take into account the concentration of silt load being transported but made it dependent upon the silt size only, whereas Blench gave relationships for sediment side and concentration as well. Simons and Albertson (1960) also modified the Lacey regime concept by considering the factors like amount of discharge, velocity distribution, slope of water surface, shape of canal cross section, suspended sediment distribution, etc.

In IBIS majority of the irrigation systems are operated on proportional water distribution basis. Sediment transport in such systems is comparatively easy to manage as canals always run at full supply discharge. The flow is always maintained more than 80% of the design discharge of the canal because the irrigation canals have been designed based on the proportional water distribution concept. In this flow control system, if flow drops below 80% of the design discharge or full supply discharge then the water distribution is does not remain equitable and sedimentation takes place in the canal (Ghumman et al., 2008). It is the general rule of canal operation in IBIS that the irrigation canals are always operated above 80% of the design discharge in order maintain equitable water distribution and to avoid sedimentation.

Almost all irrigation canals in IBIS are run-of-river irrigation canals. These irrigation canals get enormous amounts of sediment in day to day operation. Figure 9.1 presents a picture of Machai Branch Canal carrying sediment laden flow. Then depending upon the hydraulic characteristics of the canal and sediment inflow rates the process of sedimentation takes place in the canal. Generally big irrigation canals do not undergo severe sedimentation because they have relatively higher sediment transport capacity, whereas small irrigation canals face severe sedimentation problems. Figure 9.2 shows sediment deposition in Lower Machai Branch Canal.

Figure 9.1. Sediment laden flow in Machai Branch Canal

Figure 9.2. Sedimentation in a section of the Lower Machai Branch Canal

9.2. Desilting of irrigation canals

Desilting of irrigation canals is a regular exercise in IBIS which is generally practiced during the canal closure period. Every year during this period, depending upon the sedimentation in the canal, longitudinal and cross-sectional surveys of the canal are conducted to assess the sediment deposition volume, changes in canal prism, bed slope and cross-sectional areas due to sedimentation or scouring in the canal. Then depending upon the sedimentation volume the canal is desilted either manually or mechanically. Figure 9.3 presents desilting of Lower Machai Branch Canal in January 2007.

Figure 9.3. Desilting of Lower Machai Branch Canal in January 2007

Table 9.1 presents the expenditures incurred on desiltation of irrigation canals in Punjab Province of Pakistan during the five year period from 2003-2008 by the Punjab Irrigation and Power Department, Pakistan. Whereas the data on the requirement of funds for canal maintenance and desilting were not available. Owing to the gigantic irrigation network, severe sedimentation problems in the irrigation canals and lack of funds, it can be said that the required funds would be much higher than the amount actually spent. Provincial irrigation departments have established yardsticks for regular canal repair and maintenance and in the light of these yardsticks, every year in the annual budget the funds are allocated for desiltation purposes.

Table 9.1. Five year expenditures incurred on desilting of irrigation in Punjab Province, Pakistan (2003 to 2008)

S. No.	Year	No. of canals	Length desilted (km)	Quantity of silt removed (MCM)	Cost (Million US$) @ 1US$ = Rs. 60
1	2003-2004	855	8,403	8.2	2.5
2	2004-2005	536	5,408	3.5	1.1
3	2005-2006	584	5,405	2.1	1.2
4	2006-2007	704	6,098	3.3	2.2
5	2007-2008	642	5,790	5.5	4.0
	Total	3,321	31,105	23	11
	Average	664	6,221	4.5	2.2

9.3.   Effect of sediment transport on upstream controlled irrigation canals

Major effect of sediment deposition in an upstream controlled irrigation canal is the reduction in flow carrying capacity of the canal and raise in water levels. This reduction in the flow carrying capacity results in inadequate and inequitable water supply to offtaking canals and to tertiary outlets. In upstream controlled irrigation canals, with proportional water distribution system, the raise in bed levels in the head reaches causes a raise in water levels in those reaches. This results in over supply of water to tertiary outlets at the head reaches, and a reduction in supply to the outlets at downstream reaches. Generally the orifice type Adjustable Proportional Module (APM) or Adjustable Orifice Semi-Module (AOSM) outlets are installed at the head reaches, so that they have minimum effect of water level fluctuations. Even then if once water level is raised due to sedimentation, these head outlets enjoy more water round the year, until the canal is desilted during next closure period. Sometimes, sediment deposits so severely that it reaches above the sill levels of small outlets. Figure 9.4 shows an example of severe sedimentation in front of an irrigation outlet in Machai Branch Canal causing reduction in its discharge.

Inadequacy in water supply, inequity in water distribution and unreliability in irrigation flows are major existing problems in the irrigation canals of IBIS. All of these problems, especially, the former two are associated with sedimentation and delays in canal maintenance. Many researchers have reported higher delivery performance ratio (DPR) values of irrigation outlets at head reaches of irrigation canals and lower values of DPR of tail reaches outlets, where DPR is the ratio of actual discharge to design discharge of an offtake/outlet (See Ahmad et al., 1998; Khan et al., 1999; and Munir et al., 1999). These factors not only affect the irrigation canals' hydraulic performance but also cause severe impacts on crop production and sometimes even lead to complete crop failures. The tail enders famers are the major victims of poor hydraulic performance of the irrigation canals. The inadequate and unreliable water deliveries to canal tail areas lead to other environmental problems like secondary salinization along with low crop production.

Figure 9.4. Sedimentation in front of an outlet

9.4.   Effect of sediment transport on the hydraulic performance of downstream controlled irrigation canal

In automatically downstream controlled irrigation canals the effect of sediment transport is somehow different from the effects on upstream controlled irrigation canals.

As far as reduction in discharge carrying capacity of the canal and raise in water levels are concerned due to sedimentation in the canal, are the same in both types of irrigation canals. Further, in downstream controlled irrigation canals, a rise in water levels affects the auto-functioning of the hydro-mechanically operated AVIS/AVIO gates and of automatic flow controller at the system headworks. A raise in water level at the canal headworks reduces the flow releases to the canal from the automatically controlled water supply at the headworks. As soon as water supply to the canal becomes less than the outflow from the canal, the automatic functioning of water level and flow regulating AVIS/AVIO gates stops to function. The description of this phenomenon is given the following sections.

9.4.1. Effect of sediment deposition at the automatic flow releases

As the canal under study is an automatically operated irrigation canal. The discharge supplies to the canal are controlled by automatic PI discharge controllers, which respond to water levels in the canals, whereas the water levels in the canal are regulated by AVIS/AVIO gates. The hydraulic functioning of downstream control system has been discussed in Chapter 3 in detail. The flow releases at the canal headworks in such canals depend upon the difference in actual and target water levels. If the actual water level is low, the difference will be higher and more water will be supplied to the canal to meet the target water level and vice versa.

How this phenomenon takes place, is illustrated in Figure 9.5. Simulations were performed at a discharge of 26.3 m³/s, with a sediment concentration of 0.25 kg/m³ for the first 90 days and then onwards 1.00 kg/m³, with d_{50} of 0.09 mm (fine sand). It can be seen that up to 100 days of simulation, there was not a big change in the canal bed level. The bed level variations with corresponding water levels and discharge variations are given in Table 9.2. At day 110 the bed level rose to 378.76 m+MSL from 378.65 m+MSL. A corresponding increase in water levels was observed as 0.1 m, which caused a reduction in flow of about 0.5 m³/s. This situation kept on increasing and on day 140 discharge from the headworks reduced to 23.3 m³/s, which was 26.3 m³/s in the beginning of the simulation period.

Table 9.2. Bed level, water level and discharge variation at the canal head reach

Day	Concentration (kg/m³)	Bed level (m+MSL)	Water level (m+MSL)	Discharge (m³/s)
0	0.25	378.62	382.20	26.3
100	0.25	378.65	382.21	26.4
110	1.00	378.76	382.30	25.9
130	1.00	378.83	382.32	24.5
140	1.00	378.95	382.34	23.3

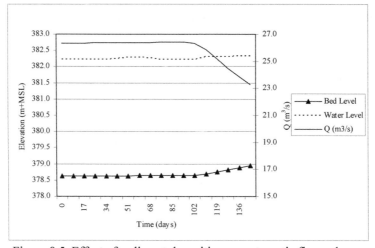

Figure 9.5. Effect of sediment deposition on automatic flow releases

How the reduction in discharge from the canal headworks affects the auto functioning of downstream control, AVIS/AVIO gates is given in the Figure 9.6. Figure 9.6 shows upstream and downstream water level variations and gate openings of one of the AVIS cross regulators of the canal. It can be seen that as the discharge reduced at the canal headworks, the water level downstream of the gate started to decline. By a decline in the water level downstream of the gate, the opening of the gate kept on increasing to maintain the downstream target water level. However, when the water level dropped to 365.14 m+MSL, the gate was opened to its maximum opening, 1.75 m, which was 0.50 m in the beginning of simulation period. The downstream water level was 365.24 m+MSL in the beginning. On the day 110, the downstream water level dropped to 365.20 m+MSL and gate opening increased to 0.80 m.

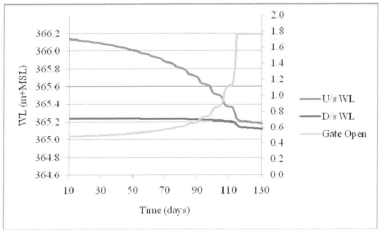

Figure 9.6. Gate response to water level variations

These are the two major phenomena which are usually encountered in downstream controlled irrigation canals. In these situations water just passes freely in the canal and the AVIS gates have no influence on flow control. The offtaking canals then cannot get their design discharges because of the unavailability of the required head to draw the required amount of discharge. Sometimes flow is controlled in the canal by fixing the float at a certain level by putting chain pulley to the float as shown in Figure 9.7.

Figure 9.7. Use of chain pulley to control flow in the Maira Branch Canal

9.4.2. Effect of secondary offtakes operation on sediment transport capacity

In the system downstream of RD 242 the secondary canals are kept closed during low water requirement periods, as discussed under CBIO. These offtakes are grouped together for periodical closing and opening purposes. A number of options are available in the CBIO model for grouping and clustering of the offtakes, which can be grouped together depending upon the operation requirement and access to the offtakes. Sometimes the canal operators prefer to close a cluster of offtakes which are close to each other, so with a minimum travelling or effort the offtakes can be opened or closed. Sometimes the offtake closure depends upon the duty area of the gate operators. All of the offtakes in the duty area of one gate operator can be kept closed or opened. Sometimes, they are grouped between two consecutive cross regulators for ease of operation. Anyhow, each option affects the sediment transport in the canal in a quite different way, which will be discussed in next section.

Simulations were performed how the pattern of offtakes closure affected the sediment transport capacity of the canal. The canal was operated at three different scenarios, at full supply discharge (FSD), and then at 50% discharge by closing first the upstream offtakes and then the downstream offtakes. It was observed that the sediment transport capacity at these three operation scenarios was different and the least was in the case of downstream reach offtakes closure. Figure 9.8 presents the simulation results of sediment transport capacity of the canal under these three canal operation scenarios. The sediment transport capacity was maximum at full supply discharge (FSD) then comparatively less under upstream offtakes closure and the least in downstream offtake closure.

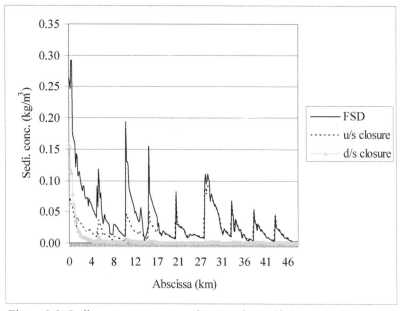

Figure 9.8. Sediment transport capacity at various offtakes operation

9.5.   Effect of different operation schemes on sediment transport

In general there are three operation schemes in the Maira Branch Canal, one is the canal operation at design discharge round the year; second is the canal operation at existing discharge; and the third is the canal operation according to CBIO. Basically the irrigation system under study has been designed for semi-demand based operations or CBIO (see Chapter 5), but sometimes water users resist operating the canal according to CBIO. Therefore, the Department of Irrigation tries to operate the canal as per design discharges. But the secondary system, due to poor maintenance or water wastage by the farmers, is operated on less than the design discharges. The effect of the three operation schemes on sediment transport has been studied by applying the model and will be discussed in the following sections.

9.5.1. Effects of design discharge

As explained before the maximum discharge at the confluence point is about 25.5 m^3/s. For simulation purposes this amount of discharge has been regarded as design discharge.

Though, from sediment transport point of view to operate an irrigation canal at its maximum or design discharge is quite desirable, as happens in general in irrigation canals of the IBIS. At the design discharge the maximum sediment is carried by the flow and is also better distributed to the offtaking canals. But the story becomes different when it comes to the irrigation canals which are designed on demand based concept. These canals have high water allowance and are designed to carry the peak discharge as well as minimum discharge, in order to meet the peak crop water requirements.

Figure 5.2 presents the existing crop water requirements of the command area downstream of RD 242. It can be seen from the Figure 5.2 that the peak crop water requirement remains for short time during a year. If the canal is operated at the maximum discharge round the year, as following the general practice in IBIS, it would cause wastage of water, which is already a scare resource in the Indus Basin. Not only water wastage, but it would cause the environmental problem of waterlogging. Table 5.1.B gives the design discharges of Machai/Maira Branch Canal system. Total discharge of the offtakes which are under demand based operation becomes 25.3 m^3/s.

As the crop water requirements are much lower (see Figure 5.2) than the maximum canal discharge, efforts are made to run the canal as close as possible to the crop water requirements. But to strictly follow the crop water requirement curve is also not favourable from sediment transport point of view. It can be seen from Figure 5.2 that sometimes discharge is so low that if the canal is operated on strictly following this curve, a lot of sedimentation would take place. Therefore some judicious canal operation rules are required, by following which the sediment deposition can be minimized while catering the CWR of the canal command area. Total volume of water needed to meet the CWR of the system downstream of RD 242 was about 470 MCM in cropping season 2007-2008. If the canal is operated at design discharge then a total volume of water of 725 MCM is delivered to the system. It becomes about 150% of the crop water requirements.

Sediment transport at design discharge with existing sediment inflow resulted in quite reasonable results under the existing sediment inflow rate. About 18,000 m^3 sediment deposition was observed under design discharge conditions in the head and middle reaches of the Maira Branch Canal. A bed level raise of 20 cm at the head reaches was observed, while it was less than 5 cm in the middle and tail reaches. Whereas the simulation results showed more sediment deposition by considering sediment inflow from the Tarbela Reservoir.

Figure 9.9 presents the sediment deposition trend in different canal reaches. A number of four different canal reaches have been selected in order to evaluate the sediment deposition pattern. The Pehure Branch reach is just downstream of the confluence and Yar Hussain (YH) reach is also located at the head of the Maira Branch Canal, whereas the Nek Nam (NN) reach is in the middle of the canal and the Qasim-2 (Q-2) reach is at the tail of the canal. It can be seen from Figure 9.9 that the canal reach downstream of the confluence observed some scouring during the low sediment concentration period, due to comparatively higher velocities there. In the other canal reaches, no scouring has been observed. The eroded material at the head reach deposited in the next reaches, where flow velocities are relatively lower. The maximum deposition took place after 120 days of canal operation, when high sediment concentration started to come into the canal (Table 7.4). Days from 120 to 240 represent the four months from June to September, which are peak sediment concentration periods.

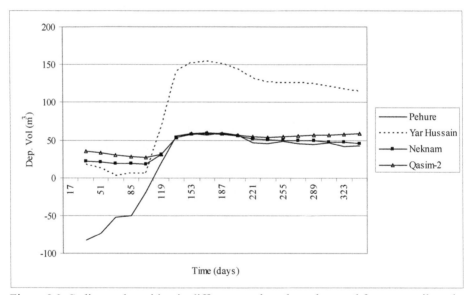

Figure 9.9. Sediment deposition in different reaches along the canal for coarse silt under design flow conditions

9.5.2. Effects of existing discharge

Almost all the time during study period, offtakes of the system downstream of the RD 242 were operated at less than the design discharge and higher than the crop water requirements. The main reasons behind this were the water users' resistance to implement the CBIO and the maintenance conditions of the secondary canals, which could not draw their design discharges due to poor maintenance. During the study period, the existing discharge at the confluence remained 75 to 90% of the design discharge, excluding the discharges of the lift irrigation schemes and additional discharges of Pehure Branch and Link Channel. However, existing discharges were not planned well, as some of the offtakes withdrew higher amounts of water, whereas the other offtakes withdrew lower amounts of water than their design discharge.

In existing discharges the canal runs almost at the same discharge round the year, which is somehow favourable for sediment transport. Figure 9.10 presents average discharges of secondary offtakes from October 2006 to September 2008. It can be seen that the discharges are lower in the months of July and August, contrary to the crop water requirements, which is maximum in these months. The high delta crops like tobacco, sugarcane and maize are major crops of this season, which need high amount of water. Sediment concentration is usually higher in these months and if the discharge is lowered in these months more sedimentation takes place in the canal.

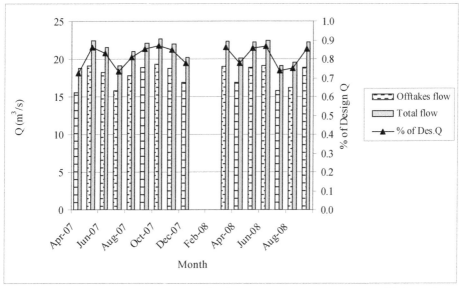

Figure 9.10. Total existing discharge in the canal including discharge withdrawn by secondary offtakes from April 2007 to September 2009

As shown in Figure 9.11 more sediment deposition took place after 120 days of canal operation i.e. in July and August. In these months, more sediments come into the canal and if offtakes diversion is reduced, it causes less discharge in the main canal and results into more sediment deposition. It can be seen from Figure 9.11 that during July-August, deposition is more under existing flow conditions as compared to the sediment deposition under design discharge conditions. The total deposition under existing flow conditions with existing sediment inflow was estimated to be 19,400 m³ in the canals downstream of RD 242 in the Lower Machai Branch Canal and in the middle reaches of Maira Branch Canal.

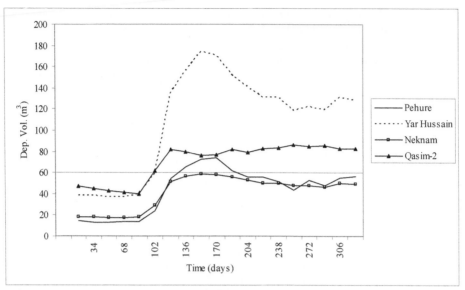

Figure 9.11. Spatio-temporal sediment deposition in the Lower Machai/Maira Branch Canal under existing discharge conditions

9.5.3. Effect of different options of CBIO on sediment transport

The crop based irrigation operation is generally prepared by using the CBIO Model (Pongput, 1998). In CBIO Model there is variety of options available to prepare the schedules. Crop water requirement is met by making different grouping and clustering of offtakes as discussed in section 9.4.2. The water supply to the canal command area depends upon the selection of the canal operation schedule, whereas schedule preparation depends upon the preference of the operator; for example water saving, environmental sustainability or sediment management. The discharge released under different schedules is different, so it directly affects the water saving and sediment transport. Generally, by supplying the maximum amount of water, more wastage of water takes place and less sediment is deposited in the canal. On the other hand, by supplying less amount of water more water can be conserved but more sediment is deposited in the canal. Therefore, the choice of CBIO schedule matters a lot in sediment management of automatically downstream controlled irrigation canals.

Figure 9.12 presents different discharge releases under four different grouping and clustering options in the CBIO. In CBIO modelling, there is a number of options available to match the water supply to the crop water requirement as close as possible. The maximum or design discharge can be supplied or can be avoided keeping in view the crop water requirements and other factors like environmental sustainability and sediment transport. Here option-1 has the least discharge supplied to the canal, nowhere during the schedule, the full supply discharge was supplied. While, option-4 has the maximum discharge supplied under CBIO and the canal is operated at design discharge or on supply based operations for more than 10 weeks.

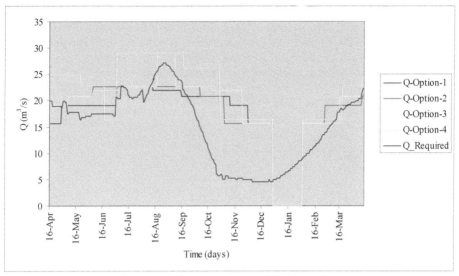

Figure 9.12. Flow releases at different options of CBIO against water requirements

Table 9.3 presents the volume of water supplied under these different CBIO options in comparison with the required volume of water to meet the crop water demands and volume under design conditions. Option-1 has the least volume supplied, which is 33% lower than the volume supplied under design conditions and option-4 has the maximum discharge. All of these options supply more water than the required amount of water due to the operation rules of CBIO (see Appendix C). Part of this over supply of water under CBIO cannot be avoided because secondary offtakes cannot be kept closed for more than one week. Therefore canal discharge cannot be reduced more than 50% of the design discharge. By comparing these options with option-4, option-1, option-2 and option-3 supply 14%, 12% and 9% less water than option 4. From supply and demand point of view, option-3 closely matches the water requirement line and was recommended for the canal operation.

Table 9.3. Volume of water supplied under different CBIO options

CBIO Options	Volume of water supplied (m³)	Difference from design volume %	Difference from maximum CBIO volume %
Option-1	551	33	14
Option-2	564	32	12
Option-3	587	29	9
Option-4	642	22	
Required volume	469	43	
Design volume	825		

Figure 9.13 presents sediment transport in Maira Branch Canal under different options of CBIO. The sediment data given in Table 7.11 has been used for simulations with d_{50} = 0.044 mm (coarse silt). Figure 9.13 shows that there is not a big difference in sediment deposition under each of these three options. Anyhow, the maximum deposition took place under option-1 whereas minimum deposition took place under option-3.

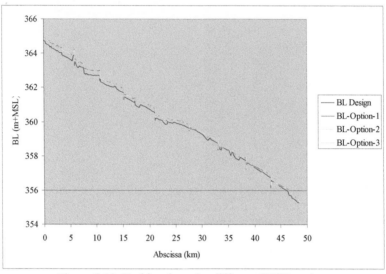

Figure 9.13. Bed levels under different CBIO options

9.6.   Management options

9.6.1. Operation under different discharge conditions

Operation under design discharges

Model simulations show that at design conditions, the sediment transport in the canal is maximum and it favours sediment transport in the canal. But this sort of operation would cause wastage of water on the one hand and environmental problems, like waterlogging on the other. The design discharges supply about 75% more water than the crop water requirements. It would not only cause the above mentioned two problems, but it would also cause maintenance problems in the secondary system. When water users do not need water, they generally close the tertiary outlets. This closure of tertiary outlets causes overtopping in the secondary canals, resulting in canal breach or berm erosion. Then it would need again high amounts of funds to repair the secondary canals. Secondly, waterlogging is already a major problem in the tail reaches of secondary canals (Munir et al., 2008). By having more water farmers tend to over irrigate the crops, which would add further pressure to the already prevailing waterlogging problem. Hence, canal operations under design discharges are not recommended in order to conserve water, to sustain the environment and to minimize the repair and maintenance costs of the secondary canals.

Operation under existing discharges

As already mentioned the supply based operation has been in practice in the IBIS since its inception. In supply based operations, the canals are supplied with full supply discharge round the year, depending upon the water availability in the rivers or in other water sources. Therefore, water users in the IBIS are used to this type of canal operation. Further, water distribution is strictly proportional in the IBIS. For proportional water distribution, the infrastructure has been developed in such a way that there is almost no flexibility in water distribution. Tertiary outlets can only get their fair share of water if the parent canal runs at full supply discharge. Therefore it is a perquisite to run the canal at full supply discharge for fair distribution of water. Generally, there is no problem in maintaining such condition, when sufficient water is available at the water source. But, in case of any shortage at the water source, this condition cannot be maintained anymore. Therefore, it is managed by keeping a part of the canals closed for a short period of time, one to two weeks, so that other canals can be operated at full supply discharge. In this way a rotation system is started and canals are operated alternatively.

Canal closure has an impression of water scarcity in the mindset of water users. Under water scarcity situations, farmers have already suffered a lot, in terms of crop failures and low crop productions. Having, such a negative impression in their minds, it is somehow difficult to convince them for canal closure for management of excess water, as proposed in CBIO. Though it is not as difficult as the water managers conceive. Anyhow, the Department of Irrigation does not bother much about it and let the water go to the water users round the year, without any implementation of CBIO. Under these circumstances, when farmers do not need water, they divert water to nearby drains, or back to the canal or they close the gated tertiary outlets. Figures 9.14.A and 9.14.B show a tertiary outlet closed by farmers and dumping water back to the canal respectively.

Figure 9.14.A. Closure of tertiary outlet by a refusal gate by farmers

Figure 9.14.B. Farmers throwing water back to the canal due to abundance of water

This refusal of water causes overtopping and berm erosion in the downstream reaches of the canal, which cause a number of maintenance problems. Figures 9.15.A and 9.15.B show a tail of a secondary canal overtopping and berm erosion due to overtopping respectively. This situation increases maintenance costs and causes damage to the nearby crops, roads and the residential areas.

Figure 9.15.A. Flooded tail of a secondary canal due to water refusal upstream

Figure 9.15.B. Berm erosion of a secondary canal due to overtopping

In order to avoid this situation, the Department of Irrigation releases a lower amount of water in the canal disregarding the CWR of the command area. This situation causes the dual problem of inequity in water distribution in the secondary canals and sedimentation in the main canal and in secondary canals. Figure 9.11 shows the sedimentation trend in the canal under existing canal flows. The sedimentation was comparatively higher, about 1,400 m^3, under existing discharges than under the design discharges. This deposition mainly took place during low canal supplies in the monsoon period, which is a high sediment inflow period. Under existing discharges about 35% more water was supplied than required. Though this amount was relatively lower than the design, but it was not planned well.

This operation technique is not efficient and leads to inequitable water distribution in the secondary canals, maintenance problems of the secondary canals and water wastage as well. Hence, this operation scheme is also not recommended for canal operation in the irrigation system under study.

Crop Based Irrigation Operations (CBIO)

Owing to the problems, like wastage of water, waterlogging and water refusal associated with the canal operation under design discharges and existing discharge, there is a need to devise some canal operation scheme, which would minimize these problems and can be efficient in sediment transport management. CBIO can be efficient in water savings and would result also better in preventing waterlogging in the canal command area. However, strictly following the CBIO causes more sediment deposition in the canal as discussed in previous sections.

In the section above on CBIO, various options under CBIO have been analysed in terms of water savings. It was found that under CBIO an amount of 20% to 30% water can be saved with different options. Whereas from sediment transport point of view, option-4 seems more efficient, but it saves the least amount of water. Whereas option-1 saves the most of the water but it causes more sedimentation. Option-3 seems more efficient as it saves more water on one hand and causes comparatively less deposition of sediment.

If Option-3 is operated with a duration of about 120 days at full supply discharge during monsoon months it changes the sediment deposition pattern in the canal. Though on average it reduces a small fraction of sediment deposition in the canal, about 5%. But, at head reaches, it reduces about 10%, in the middle reaches about 4% and at the tail reaches it increases about 1%. Figures 9.16.A and 9.16.B show the sediment deposition trend under both of these operation schemes. Various canal reaches from head to tail have been compared for the sediment deposition under normal CBIO option-3 and with a full supply extended (FSE) period with Option-3.

In these simulations the canal was operated at full supply discharge from day 119 to 238 as given in Table 9.4. Pehure Branch and Link Channel are remained open throughout the year, so they are not given in the scheduling table. It can be seen from the Figures 9.16.A and 9.16.B that there is a clear difference in sediment deposition in these days, especially in the head and middle canal reaches. In the canal head reaches Pehure, Yar Hussain (YH) and Ghazikot (GK) the sediment deposition clearly reduced under full supply extended period. The situation is somehow different in the middle and tail canal reaches. The sediment deposition in the middle reach Yaqubi (YB) is more under full supply extended period, contrary to the head reaches, where it is reduced under full supply extended period. Similarly, in tail reach Qasim-2, the sediment deposition under full supply extended period is not much different from the sediment deposition without full supply extended period, but the trend is quite different. Without the full supply extended period, the deposition pattern is uneven, whereas it is quite regular under the full supply discharge period. By following this scheme about 9% more water is supplies than water supplied under Option-3.

Sediment deposition patter as shown in Figures 9.16.A and 9.16.B show that by operating the canal at full supply discharge during peak sediment inflow period not only reduces sediment deposition but also helps in better distribution of sediment along the canal. It does not concentrate in the head reaches of the canal as generally happens during low flow periods and hence reduces the risk of flow reduction in the upstream reaches of the canal. Depending upon sediment concentration in the flow the deposited material during low flows can also be eroded and distributed to offtaking canals during full supply discharge. It shows that the demand based or semi-demand based irrigation canals having sedimentation problem should be operated at full supply discharge for some time during the canal operation, depending upon the rate of sediment inflow and deposition.

Table 9.4. Canal operation plan for sediment management

Date	Day	Sb Mn	OIB	DD	YHM	Sd Mn	GKM	NNM	Su Mn	YM	Sc Mn	DM	G1M	G2M	Q1M	Q2M	TM	PSD	CD
01-Feb	0	O	C	C	O	O	O	O	O	C	O	C	C	O	O	O	O	O	C
08-Feb	7	C	O	O	C	C	C	C	C	O	C	O	O	C	C	C	C	C	O
15-Feb	14	O	C	C	O	O	O	O	O	C	O	C	C	O	O	O	O	O	C
22-Feb	21	C	O	O	C	C	C	C	C	O	C	O	O	C	C	C	C	C	O
01-Mar	28	O	C	O	O	O	O	O	O	O	O	C	O	O	O	O	O	O	C
08-Mar	35	O	O	O	O	O	C	O	O	C	O	O	C	C	O	O	C	C	O
15-Mar	42	C	O	C	C	C	O	C	C	O	C	O	O	O	C	C	O	O	O
22-Mar	49	O	C	O	O	O	O	O	O	O	O	C	O	O	O	O	O	O	C
29-Mar	56	O	O	O	O	O	C	O	O	C	O	O	C	C	O	O	C	C	O
05-Apr	63	C	O	C	C	C	O	C	C	O	C	O	O	O	C	C	O	O	O
16-Apr	70	O	C	C	O	O	O	O	O	C	O	C	C	O	O	O	O	O	C
23-Apr	77	C	O	O	C	C	C	C	C	O	C	O	O	C	C	C	C	C	O
30-Apr	84	O	C	O	O	O	O	O	O	O	O	C	O	O	O	O	O	O	C
07-May	91	O	O	O	O	O	C	O	O	C	O	O	C	C	O	O	C	C	O
14-May	98	C	O	C	C	C	O	C	C	O	C	O	O	O	C	C	O	O	O
21-May	105	O	C	O	O	O	O	O	O	O	O	C	O	O	O	O	O	O	C
28-May	112	O	O	O	O	O	C	O	O	C	O	O	C	C	O	O	C	C	O
04-Jun	119	O	O	O	O	O	O	O	O	O	O	O	O	O	O	O	O	O	O
11-Jun	126	O	O	O	O	O	O	O	O	O	O	O	O	O	O	O	O	O	O
18-Jun	133	O	O	O	O	O	O	O	O	O	O	O	O	O	O	O	O	O	O
25-Jun	140	O	O	O	O	O	O	O	O	O	O	O	O	O	O	O	O	O	O
02-Jul	147	O	O	O	O	O	O	O	O	O	O	O	O	O	O	O	O	O	O
09-Jul	154	O	O	O	O	O	O	O	O	O	O	O	O	O	O	O	O	O	O
16-Jul	161	O	O	O	O	O	O	O	O	O	O	O	O	O	O	O	O	O	O
23-Jul	168	O	O	O	O	O	O	O	O	O	O	O	O	O	O	O	O	O	O
30-Jul	175	O	O	O	O	O	O	O	O	O	O	O	O	O	O	O	O	O	O
06-Aug	182	O	O	O	O	O	O	O	O	O	O	O	O	O	O	O	O	O	O
13-Aug	189	O	O	O	O	O	O	O	O	O	O	O	O	O	O	O	O	O	O
20-Aug	196	O	O	O	O	O	O	O	O	O	O	O	O	O	O	O	O	O	O
27-Aug	203	O	O	O	O	O	O	O	O	O	O	O	O	O	O	O	O	O	O
03-Sep	210	O	O	O	O	O	O	O	O	O	O	O	O	O	O	O	O	O	O
10-Sep	217	O	O	O	O	O	O	O	O	O	O	O	O	O	O	O	O	O	O
17-Sep	224	O	O	O	O	O	O	O	O	O	O	O	O	O	O	O	O	O	O
24-Sep	231	O	O	O	O	O	O	O	O	O	O	O	O	O	O	O	O	O	O
01-Oct	238	O	O	C	O	O	C	O	O	O	O	O	O	C	C	O	C	O	O
08-Oct	245	C	O	O	C	O	O	C	C	O	C	C	O	O	O	C	O	O	O
15-Oct	252	O	C	O	O	C	O	O	O	O	O	O	O	O	O	O	O	O	C
22-Oct	259	O	O	O	O	O	O	O	O	C	O	C	O	O	O	O	O	C	O
29-Oct	266	O	O	C	O	O	C	O	O	O	O	O	O	C	C	O	C	O	O
05-Nov	273	C	O	O	C	O	O	C	C	O	C	C	O	O	O	C	O	O	O
12-Nov	280	O	C	O	O	O	O	O	O	O	O	C	O	O	O	O	O	O	C
19-Nov	287	O	O	O	O	O	C	O	O	C	O	O	C	C	O	O	C	C	O
26-Nov	294	C	O	C	C	C	O	C	C	O	C	O	O	O	C	C	O	O	O
03-Dec	301	O	C	C	O	O	O	O	O	C	O	C	C	O	O	O	O	O	C
10-Dec	308	C	O	O	C	C	C	C	C	O	C	O	O	C	C	C	C	C	O
17-Dec	315	O	C	C	O	O	O	O	O	C	O	C	C	O	O	O	O	O	C
24-Dec	322	C	O	O	C	C	C	C	C	O	C	O	O	C	C	C	C	C	O

In the table above: O = open, C = closed, Sb Mn = Sarbandi Minor, OIB = Old Indus Branch, DD = Dagi Distributary, YHM = Yar Hussain Minor, Sd Mn = Sadri Minor, GKM = Ghazikot Minor, NNM = Nek Nam Minor, Su Mn = Sudher Minor, YM = Yaqubi Minor, Sc Mn = Sardchina Minor, DlM = Daulat Minor, G1M = Gumbat-1 Minor, G2M = Gumbat-2 Minor, Q1M = Qasim-1 Minor, Q2M = Qasim-2 Minor, TM = Toru Minor, PSD = Pirsabaq Distribuatry, CD = Chowki Distributary

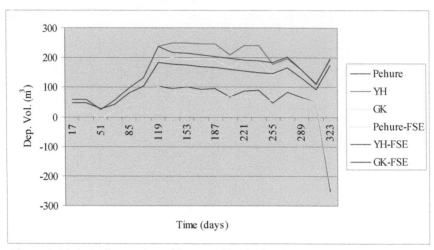

Figure 9.16.A. Sediment deposition trend in the head reaches of the canal during implementation of CBIO with and without full supply discharge extended period

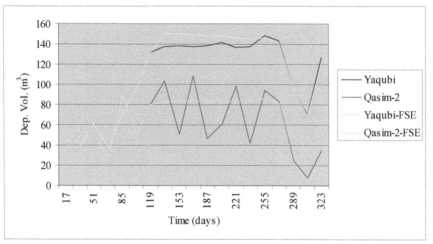

Figure 9.16.B. Sediment deposition trend in the middle and the tail reaches of the canal with and without full supply discharge extended period

9.6.2. Sediment transport under CBIO with increased sediment discharge from the Tarbela Reservoir

Another scenario has been simulated to find the suitable CBIO option under increased sediment discharge from the Tarbela Reservoir. The CBIO option-3 with and without full supply extended period has been simulated. Under this scenario the Option-3 of CBIO with full supply extended period has proved to be quite efficient. As the total sediment deposition under CBIO option-3 was found to be 85,000 m^3 for coarse silt, whereas the sediment deposition under this option with full supply extended period was found to be 60,000 m³.

These results show that the sediment deposition can be reduced up to 27% by operation the canal at CBIO with a full supply extended period. The canal is operated for about 120 days at full supply discharge, mainly during the monsoon months. During this time of the year, the crop water requirements are also high due to cultivation of high delta and high value crops like tobacco, maize and sugarcane. Sediment deposition in PHLC and the canals downstream of RD 242 is shown in Figure 9.17.A and Figure 9.17.B respectively. The Figures 9.16.A and 9.16.B show that the sediment deposition under CBIO-FSE has been decreased significantly, particularly at head reaches, which is good for canal operations due to low hindrance to the water flow, resulting in adequate water supplies to the system.

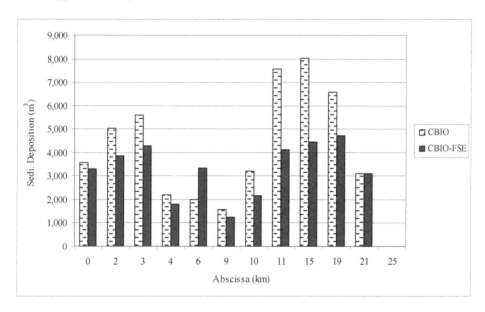

Figure 9.17.A. Sediment deposition in different reaches of PHLC under CBIO and CBIO-FSE

The overall sediment deposition volumes in the PHLC and in the canals downstream of RD 242 (the Lower Machai Branch Canal and the Maira Branch Canal) under both types of canal operations are given in Table 9.5.

Table 9.5. Sediment deposition volumes in PHLC and in the canals downstream of RD 242

Canal	Sediment deposition (m³)	
	CBIO	CBIO-FSE
PHLC	43,500	34,900
d.s. of RD 242	41,400	27,800
Total	84,900	62,700

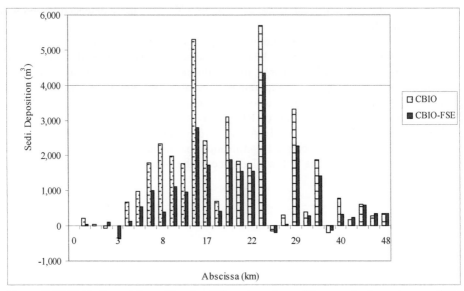

Figure 9.17.B. Sediment deposition in different reaches of the canals downstream of RD 242 under CBIO and CBIO-FSE

9.6.3. Target water level and sediment transport

As automatically downstream controlled irrigation canals are operated on the basis of a target water level at the downstream side of the cross regulators. These canals are usually equipped with automatic flow controllers at the headworks. These controllers can be downstream control gates or PI controllers or of some other types. This system of flow control works fine in sediment free environment. The canals receiving sediment laden flow need some management for smooth canal operation. In case of sediment deposition in the head reaches, the water level rises up and causes reduction in the flow releases from the automatic flow controllers. The objective of the PI controller is generally to meet the target water level, not the target discharge. Therefore, the target water level is maintained, but at low flows due to sedimentation in the canal prism as shown in Figure 9.18.

There are two options to deal with this situation, first the intermittent dredging of the canal, in order to lower the canal bed levels and second is to increase the target water level, so that more flow can be supplied in case of a raise in water level, depending upon the freeboard available in the canal. Simulations have been performed to see how this phenomenon would work. Figure 9.18 shows that at the target water level of 366.64 m+MSL the discharge started to reduce after 280 days of canal operation. The target water level was increased to 366.74, an increase of 0.10 m, on the day 90 of canal operation. The required discharge was maintained until the end of the required canal operational period. In downstream control system, there is a limitation in setting the maximum target water level that it should not be higher than the allowed maximum water level at the upstream of the downstream AVIS cross regulator. Otherwise, in case of gate closure, there would be overtopping at the gate, causing damage to the structure.

So, when a reduction in the flow is observed from the discharge releases data at the PI controller, the target water level can be increased to some extent depending upon the sediment deposition, conveyance capacity of the canal and maximum allowed water level at the downstream AVIS cross regulator.

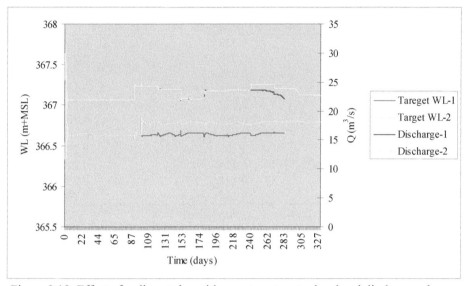

Figure 9.18. Effect of sediment deposition on target water level and discharge releases

9.6.4. Decrement setting and AVIS gates' response

The decrement of AVIS gates is the key parameter which controls hydraulic functioning of the gate. The decrement is the difference in water levels at full supply discharge and at zero flow discharge. At full supply discharge the gate is usually open at its maximum opening whereas at zero flow the gate is almost closed. The setting of the decrement then matters in defining the gate's hydraulic behaviour in response of any change in the flow. The general criterion for setting the decrement of an AVIS/AVIO gate is based on the float radius of the gate. It is usually $0.032r$ to $0.05r$, where r stands for the float radius. This range may also go to $0.10r$ depending upon the requirements of canal operation. The influencing factors in selecting a decrement are generally the stability of the system, reaction to water levels changes and the response times, etc. The gate manufacturing companies generally give the range of the decrement with the gates. The recommended values of the decrement for high head AVIS gates are given in Table 9.6 (Baume, 2005).

Generally larger decrements are more responsive to flow changes. A small change in flow or water level causes the gate to set at new position earlier. The gates with small decrements are comparatively less responsive to flow changes. So a small change is water level does not cause the gate to move immediately at new positions. This factor can play an important role in downstream controlled canal operations with sediment deposition problem. The raise in water level due to sedimentation affects the gate opening, which is function of the decrement. In case of large decrement the gate sets at new position immediately with reduced opening and causes reduction in water supply to the downstream canal reaches. Whereas in case of smaller decrement, the gate does not respond so quickly. In this case the gate opening is not reduced immediately and the downstream canal reaches can enjoy the required amount of water for longer period.

Table 9.6. Minimum and maximum decrement values for high head AVIS gates

Gate index	Minimum decrement (m)	Maximum decrement (m)
56/106	0.029	0.058
71/132	0.036	0.072
90/170	0.045	0.090
110/212	0.058	0.115
140/265	0.072	0.143
160/300	0.080	0.160
180/335	0.090	0.179
200/375	0.101	0.202
220/425	0.114	0.227
250/475	0.128	0.256
280/530	0.144	0.288

Figure 9.19 presents the effect of decrement setting on discharge releases at the system headworks. It can be seen that due to sediment deposition and water level rise the larger decrement caused the reduction in discharge supplies earlier as compared to the smaller decrement. Figure 9.20 shows the gate opening of cross regular No. 6 at two different decrements. The decrement values of 0.101 m and 0.202 m were used. It can be seen that gate opening under decrement value of 0.101 is somewhat stable and remained almost same during the simulation period, which was two months. Whereas the gate opening under decrement value of 0.202 started increasing after day 14 of canal operation. The increase in the gate opening was due to the fact that sediment deposition in the canal raised the water level, which ultimately reduced the gate opening. Then the raise in the water level started to reduce the flow releases from the system headworks. Then ultimately, due to decreased flow upstream the gate continued to increase the opening in order to maintain the downstream target water level.

Hence it can be said on the basis of these simulation results that a small decrement slows down the response of the system and is better for canals having sediment deposition problems. A small decrement would then ensure the required supplies from the canal headworks.

9.6.5. Grouping and clustering of offtakes

As mentioned earlier, the grouping and clustering of offtakes affects the sediment transport capacity in the canal. The grouping of the offtakes from the tail cluster affects the sediment transport capacity of the canal, the most. Hence, it is not recommended to group offtakes for closure at the tail reaches of the canal. The grouping of offtakes all along the canal favours sediment transport in the canal. Hence, it is recommended that under CBIO, the secondary offtakes would be grouped for opening and closing all along the canal instead from the reaches between two consecutive cross-regulators or from the tail.

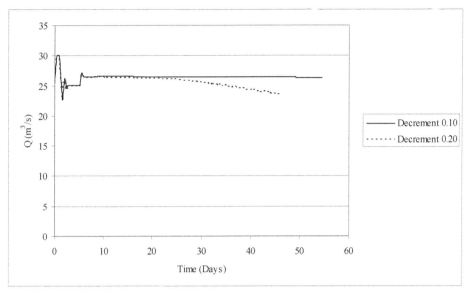

Figure 9.19. Effect of decrement setting on flow releases at the system headworks

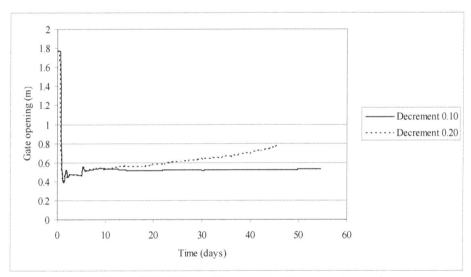

Figure 9.20. Effect of decrement setting on gate opening

9.6.6. Offtakes close to cross regulators

In upstream control system the head regulators of secondary offtakes are installed close to cross regulators in the upstream direction in the canal. In this way a proper head above the offtake's sill level can be maintained and the offtake can draw its fair share of water. The required head above the crest of the offtakes' head regulators is maintained by adjusting these cross regulators. But this phenomenon does not support much the withdrawal of sediment by the secondary offtakes. In automatically downstream controlled canals, the location of the head regulators of the secondary offtakes is kept close to the cross regulators in the downstream direction in the canal. In this way the discharge in the secondary offtake does not fluctuate much due to less variations in the water level downstream of the cross regulators in the main canal.

This phenomenon is quite favourable for sediment diversion to the secondary canals. A lot of mixing of flow and sediments takes place at the downstream side of the cross regulators. Then it takes some distance for settling again. It has been observed during the field measurements that the offtakes close to cross regulators withdrew more share of sediment as compared to offtakes far from the cross regulators. For example, the Dagi Distributary, Sadri Minor, Sudher Minor, Sardchina Minor, Gumbat-II Minor and Qasim-1 Minor have the more sediment withdrawal efficiency than the other secondary offtakes (see Chapter 6). The average sediment withdrawal efficiency of these offtakes was 70%, which was about 14% higher than the other offtakes. Hence, during the design of the downstream controlled irrigation canals, the location of secondary offtakes should be kept close to the cross-regulators in the downstream side, in order to distribute fair share of sediment to the offtaking canals.

10. Evaluation

10.1. Sedimentation in Tarbela Reservoir and PHLC

It has been estimated that the average annual sediment inflow to the Tarbela Reservoir has been about 203 million tonnes since the construction of the dam and that 95% of this settles into the reservoir. Due to this massive settlement of the sediments the reservoirs has lost about 32% of its gross storage capacity. The high sediment inflow to the reservoir, the reservoir operation policy and the sedimentation pattern in the reservoir has created a huge sediment delta in the reservoir. The delta was moving towards the dam at a rate of approximately 0.5 km per year up to 2003. Since then the delta's forward movement is almost stagnant. It might be due to the adoption of reservoir operation policy of raising minimum operational water level to a height of 417.3 m+MSL from 396.2 m+MSL. The sediment delta has reached now at a distance of 10.5 km from the dam.

If the delta continues to move forward or is disturbed by an earthquake action it may completely clog the tunnel inlets. A number of studies were done by various organizations to control sedimentation in the reservoir. The important one is a study conducted by HR Wallingford in 1998. They proposed four measures to control sediment transport in the reservoir i.e. to focus on an operational policy which would keep the risks to the tunnels low, the construction of an underwater dyke to protect intakes and the construction of low level flushing system, with a capacity of 7,000 m^3/s, which would maintain quasi stable sedimentation conditions with a reservoir capacity of approximately 55% of the original value. Later Haq and Abbas (2006) evaluated these suggestions and recommended only reservoir operational policy amongst these four options.

It can be said that at the moment there is not any particular sediment management plan except the reservoir operation policy. This situation implies that the sedimentation will continue in the reservoir and the delta may also move closer to the tunnels. This will then increase the risk of sediment discharge to the intake tunnels. Therefore it is highly likely that in the near future PHLC will get sediment laden flows from the reservoir and the sediment content in the flow will increase with the passage of time.

10.2. Hydrodynamic behaviour of the irrigation canals

The irrigation system has been divided in two parts: upstream of RD 242, consisting of Machai Branch Canal and downstream of RD 242 consisting of Lower Machai Branch Canal, Maira Branch Canal and PHLC. The Lower Machai Branch Canal and the Maira Branch Canal can been considered as one canal for simplicity and analysis purpose because the Lower Machai Branch Canal is 3.5 km long and ends into Maira Branch Canal, which is 45 km long. There is no difference in flow control system and operation scheme in Lower Machai and Maira Branch Canals. Both have the same type of infrastructure, the same flow regulating mechanism and same semi-demand based operation and one ends into the other. The Machai Branch Canal, which is upstream of RD 242 is a supply based canal with fixed canal operation, whereas canals downstream of RD 242 are downstream controlled with semi-demand based operation.

In order to understand the canals' hydrodynamic behaviour under different flow conditions the flow simulations were performed by using a 1-D hydrodynamic model, SIC. The model was calibrated and validated with the field data before using it for flow simulations. The simulations were performed for design discharge or full supply discharge and for minimum discharge, because in the winter months the Machai Branch Canal receives less water from the Swat River. The simulation results depict that under design conditions all the three canals satisfactorily pass the design discharge. Under low flow conditions the water levels in Machai Branch Canal went down and a number of offtakes in Machai Branch Canal could not get their design discharge. To achieve the design discharges in the deficient offtakes the operation of the respective cross regulators has been proposed and new gate openings have been calculated.

The hydrodynamic performance of PHLC, Lower Machai and Maira Branch Canal was evaluated for automatic discharge controllers and for semi-demand based operations CBIO. It was found that a proportional gain coefficient, K_p = 2.5 and integral time coefficient, T_i = 3,000 seconds led to comparatively stable system operations for these canals. It was also found that the amount and location of discharge refusal in the canal affected response times and stability of flows. Offtakes closed all along the canal resulted in shorter response times and early stability as compared to other options of grouping at middle or tail. In case of small amounts of discharge refusal at secondary canals the main canal achieved new steady state conditions earlier and vice versa. Four discharges under CBIO were tested and it was found that the gradual increase or decrease in the withdrawals favoured smooth system operations and achieved new equilibrium conditions earlier as compared to large variations in discharge. The hydro-mechanical cross regulators responded quite efficiently to water level changes and settled at new positions without any oscillations at the given amounts of discharges under the CBIO. The results depict that by adopting the developed guidelines the operational performance and sustainability of the study irrigation canals can be improved.

10.3. Flow and sediment transport in the irrigation canals

The water supply and water distribution in the irrigation canals under study were assessed from the operation data of about two years from October 2006 to September 2008. The average supplies to Machai Branch Canal and PHLC were 28.8 m³/s and 12.4 m³/s respectively, whereas the water available at the Lower Machai Branch Canal head (or RD 242) was 12.4 m³/s. These discharges were much lower than the design discharges, which were 66.7 m³/s, 31.9 m³/s and 28.3 m³/s for Machai Branch Canal, Lower Machai Branch Canal and PHLC. The low discharges were mainly due to low water availability for Machai Branch Canal whereas for Lower Machai Branch Canal and PHLC there was less demand for water in the command area. Similarly the water distribution to the secondary offtakes was analyzed and it was found that almost all of the secondary offtakes had DPR value less than 1 in Machai Branch Canal and some offtakes in Lower Machai Branch Canal and Maira Branch Canal. Low DPR values in Machai Branch canal have the similar reason of low water availability in the canal and for Lower Machai Branch Canal and Maira Branch Canal it was due to lower water demands in the secondary canals.

Sediment transport in the Machai Branch Canal, Lower Machai Branch Canal and Maira Branch Canal was assessed for more than one and half year from February 2007 to August 2008. The average sediment inflow at Machai Branch Canal head and at Lower Machai Branch head (or at RD 242) was 0.226 kg/m^3 and 0.17 kg/m^3 respectively with high sediment inflow in summer and low sediment inflow in winter. Coarse silt was the dominant class in the sediments consisting of 60-70% of the total sediment concentration in the flow, whereas sand and clay were 10-15% and 15-25% respectively.

The sediment transport in the canal was assessed by conducting three water and sediment mass balance studies in which all the water and sediment inflows and outflows were measured. These measurements were conducted under high and low water and sediment discharges in order to know the sediment transport behaviour under different water and sediment discharge conditions. The measurements were conducted in July 2007, December 2007 and then in July 2008. In July 2007 the sediment concentration was 0.18 kg/m^3 with a water discharge of 44.8 m^3/s, which was 67% of the design discharge. The total load that entered into the canal was 681 tonnes per day and 37% of this load was deposited in the canals, whereas rest of the sediment load was withdrawn by the offtaking canals. In the second measurement, which was conducted in December 2007 the sediment concentration was 0.10 kg/m^3 and water discharge was 25.5 m^3/s, which was 38% of the design discharge. It was found that about 60% of the sediment was deposited in the canals. The third measurement was conducted on 22 July 2008, when the sediment concentration was 0.26 kg/m^3 with a water discharge of 28.7 m^3/s, which was 45% of the design discharge. It was found that a total of 654 tonnes per day sediment entered into the canals and about 53% of this sediment load was deposited in the canals. Based on these overall results it can be said that the sediment transport in irrigation canals is highly dependent upon the amount of flow available. Any decrease in flow leads to more sedimentation in the canals.

However if the sedimentation under supply and demand operation is compared then supply based operation are more efficient. During field measurements the overall average water discharge remained about 51% of the design discharge in the supply based irrigation canal and about 22% of the incoming sediment was deposited. Whereas in demand based part the average water flow remained about 65% of the design discharge during these measurements and 54% of the incoming sediment was deposited in these canals. Further, the offtakes in the supply based part have more sediment withdrawal efficiencies. These results imply that the demand based irrigation canals have more tendency of sediment deposition than the supply based canals.

The sediment size distribution was also assessed along the canals and the coarser particles were found in the head reaches and finer particles were found at the tail reaches on average. The sediment deposition was also assessed in the canals and was found that about 19,300 m^3 of the sediment deposited in the Lower Machai Branch Canal and the Maira Branch Canal during one year canal operation from February 2007 to December 2007. These results reveal that under the existing rate of sediment inflow there is not a big problem of sedimentation in the canals under study.

10.4. Sediment transport modelling

Though under the existing sediment inflow, the sedimentation is not a major problem in the canals under study. But with inception of sediment discharge from the Tarbela Reservoir, these canals can undergo to serious sedimentation problems. Therefore to evaluate the sediment transport under increased sediment inflow the sediment transport model, SIC, was used after calibration and validation to simulate a variety of scenarios of water management and sediment inflow under Tarbela influence. Before performing simulations with increased sediment discharge, the amount and type of expected sediment discharge from the reservoir have been defined on the basis of the data and studies available on sediment transport in Tarbela Reservoir. Then on the basis of the envisaged sediment discharge the simulations were performed under three possible water flow conditions in the canals, which were full supply discharge (FSD) or design discharge, existing discharge and the CBIO.

Not only for increased sediment inflow but the simulations were also performed for determining the sediment behaviour under different operation schemes in the irrigation canals under study. First the simulations were performed under the existing condition of sediment inflow in Lower Machai Branch Canal and the Maira Branch Canal. In these simulations the sediment transport was assessed under the three canal operation schemes. A total deposition of respectively 19,400 m^3, 18,000 m^3 and 22,400 m^3 was found. The average discharges released under existing discharge and full supply discharges were 18.6 m^3/s and 25.2 m^3/s, whereas under CBIO the discharge varied from 14 m^3/s to 25.2 m^3/s, depending upon the CBIO schedule and crop water requirements. The least sedimentation took place under high discharges and the high deposition took place under low discharges.

In order to determine the effect of flow variation on sediment transport the simulations were performed to assess the sediment transport capacities of PHLC, Lower Machai Branch Canal and Maira Branch Canal under different flow conditions. These simulations were performed at maximum discharge, which was the full supply discharge and at minimum discharge, which was 50% of the full supply discharge for different sediment sizes consisting of fine sand, coarse silt, medium silt and fine silt. Sediment transport capacity for fine sand at full supply discharge was 1.0 kg/m^3 to 0.7 kg/m^3, whereas at 50% of FSD it was 0.1 kg/m^3 to 0.15 kg/m^3 in PHLC. Similarly in Maira Branch Canal the sediment transport capacity for fine sand was almost nil, whereas for coarse silt it was 0.4 to 0.1 kg/m^3 along the canal at FSD whereas at 50% discharge it reduced to 0.05 kg/m^3 to 0.01 kg/m^3.

On the basis of the data and studies available on sediment transport in Tarbela Reservoir it can be said that the sediment discharge from the reservoir could be as low as 0.005 kg/m^3 and much higher than 1.00 kg/m^3 with different median sediment sizes from fine sand to fine silt, in different time periods of the year depending upon sediment inflow to the reservoir, water levels in the reservoir and the discharge releases to the PHLC and other tunnels. Anyhow, for modelling purposes sediment discharge from 0.005 kg/m^3 to 1.00 kg/m^3 with median particle sizes from fine sand to fine silt was considered.

It was determined through modelling that with this increased sediment discharge the canals under study could undergo to serious sedimentation problems, particularly in case of coarser particles like fine sand. It was observed that in case of fine sand, most of the sediments were deposited in the upstream canal PHLC, particularly under existing flow conditions and under CBIO. Under these two types of canal operations the fine sand would be deposited mainly at head reaches. The deposition at head reaches then would reduce the canal conveyance capacity and cause a raise in water level. This raise in water levels would reduce flow releases from the automatic flow controllers. Resultantly a negative water balance would develop in the canal because the inflow of water would become less than the outflow. In this situation the model simulations were interrupted and terminated. This situation occurred during the simulations with fine sand under existing flow conditions and under CBIO. The simulations were interrupted after about 188 days of simulations due to the above mentioned reasons. Total deposition was found to be about 81,600 m^3 and 71,000 m^3 under existing flow conditions and under CBIO respectively. In this situation additional measures are required to remove the deposited sediments in addition to the traditional sediment removal practice, which is conducted in January every year. The simulations under full supply discharge with fine sand continued for one complete operational year (330 days) and a 67,000 m^3 sediment volume was deposited. In case of fine silt the simulations were continued for a complete operation year without any interruption. The total sediment deposition with coarse silt under full supply discharge, exiting discharge and under CBIO was about 43,000 m^3, 76,000 m^3 and 85,000 m^3 respectively. Similarly the sediment deposition for fine silt was respectively about 10,000 m^3, 20,000 m^3 and 22,000 m^3. In case of fine sand the major portion was deposited in PHLC and for coarse silt and fine silt the major portion of sediments was deposited in Lower Machai Branch Canal and the Maira Branch Canal.

In previous simulations the sediment inflow was considered from a single source, either from Machai Branch Canal or from Tarbela Reservoir, whereas at the confluence point the water and sediment come from two sources. The SIC model does not accept sediment inflow from more than one point in unsteady state simulations. Therefore in the first step it was assessed that how much sediment concentration was arriving at confluence from the PHLC head and then the sediment coming from Machai Branch Canal was added into it to get the new combined concentration. As the coarse silt is expected to be a major class of sediments from the Tarbela Reservoir and also sediments coming from Machai Branch Canal come into the same category, therefore this class of sediment was used for further analysis. It was found that the coarse silt would arrive at the confluence or in Lower Machai Branch Canal at an amount between 0.05 kg/m^3 to 0.9 kg/m^3 from PHLC head at different flow conditions from full supply discharge to 50% of the full supply discharge. Then in combination with the sediment coming from Machai Branch Canal, the combined concentration at the confluence would range from 0.2 kg/m^3 to 0.88 kg/m^3 for coarse silt.

On the basis of this combined sediment concentration the simulations were performed for Lower Machai Branch Canal and Maira Branch Canal in order to assess its effects on the hydrodynamic behaviour of these canals. It was found that coarse silt will be deposited respectively about 43,000 m^3, 51,600 m^3 and 50,900 m^3. The sedimentation of coarse silt would mainly take place at the head reaches.

The effect of sedimentation on the hydrodynamic behaviour of the canal was also assessed and how the raise in bed level would affect the flow deliveries from the automatic discharge controllers. It was observed that under existing discharges with coarse silt there would be no significant reduction in flow from the controller but in case of fine sand the flow would start to reduce after 217 days of canal operation and then the discharge would continue to drop gradually. The effect of this reduction in flow would also affect the behaviour of AVIS/AVIO cross regulators. In case of significant drop in water levels in the canal these gates cannot control flow anymore.

10.5. Development of downstream control component in SETRIC Model

The task of modelling sediment transport in demand based, downstream controlled irrigation canals with an automatic flow regulation system is too complicated and creates much difficulties during the modelling process. A fully hydrodynamic model consisting of an unsteady flow component coupled with an automatic flow regulation module and also having a sediment transport component is capable for simulating sediment transport in such irrigation canals. Usually steady state models are not applied in demand based systems because the existing steady state models are not able to adjust the flow automatically. If unsteady state models are used then they need a regulation module and only a few models have such regulation modules.

To overcome this complication a simple steady state model SETRIC was improved to simulate sediment transport in demand based downstream controlled irrigation canals. In this model the downstream control component was added and the option of automatic adjustment of the hydrograph was incorporated. Then the SETRIC Model was applied to Lower Machai Branch Canal and Maira Branch Canal for flow and sediment transport simulations. The model was calibrated and validated with the field data. Three different scenarios of flow with existing sediment inflow rate were simulated. The results were in good comparison with the field measurements and with results from the SIC Model. It shows that the SETRIC Model can be applied to downstream controlled irrigation canals for flow and sediment transport modelling.

10.6. Managing sediment transport

Historically sediment management has been part of irrigation management since the inception of the irrigation system in IBIS. To control sedimentation, the canals are designed generally on Lacey's regime concept and then they are always operated at full supply discharge in order to inhibit sediment deposition in the canals. Even then sediment deposits in the irrigation canals which is then removed physically on annual basis. Majority of the irrigation canals in IBIS are supply based with upstream control but now moving gradually to demand based operation. In upstream controlled irrigation canals the sedimentation affects generally the adequacy in water deliveries and equity in water distribution, whereas in demand based automatically controlled irrigation canals the problem is somehow different. The automatic flow controllers at the canal head responds to water levels. A raise in water level due to sedimentation reduces flow from the controller. Then correspondingly the behaviour of AVIS/AVIO gates is affected and they cannot control flow in the canal.

It was found during field measurements and model simulations that the canal operation at full supply discharge favoured the sediment transport. But under this canal operation the total water delivered to the irrigation system downstream of RD 242 was about 725 MCM, which was 50% higher than the crop water requirements of the area. The total sediment deposition under full supply discharges with existing sediment inflow was found to be 18,000 m³. The sediment deposition under existing discharges was evaluated. The total water delivered under existing discharge was about 35% higher than the CWR. But it was quite irregular because in peak requirement periods the discharge remained low whereas in low requirement periods it remained generally high. The total sediment deposition under existing discharges was about 19,400 m³. In summer months more sediments came to the canals and due to low flows under existing discharge conditions, more sedimentation took place in middle reaches of Maira Branch Canal.

In comparison with canal operation under full supply discharge or existing discharge, the CBIO is a quite effective technique to balance the water supply with CWR. Even within CBIO there are many options of grouping and clustering of offtakes for water supply to meet the CWR as close as possible. The sediment transport was evaluated for four different such options of CBIO. Flow supplied under each of the CBIO schemes was different. Option 1 had the least amount of water supplied, 551 MCM, whereas Option 4 had the maximum amount of water supplied which was 642 MCM. It was found that there was only ± 5% difference in total volume of sediment deposition under these options, which was 22,000 m³. The maximum deposition took place under Option 1 and minimum deposition took place under Option 3.

As far as suitable canal operations are concerned the canal operation under design discharges is not recommended because it causes about 50% more water supplies than the CWR, which results in high water losses and would also cause waterlogging in the canal command area. Operation under existing discharges is also not recommended because it causes many maintenance problems in the secondary offtakes due to over or under supplies and farmers' manipulation with canal structures. The water deliveries to the secondary canals were not planned so when there was over supply or under supply in the secondary offtakes the farmers either closed their outlets and let the water back into the canal or they just blocked the canal to get high amounts of water. The Option-3 in CBIO can be a better option for canal operations with certain modifications. It somehow favours sediment transport if water supply during summer months, June to August, is kept at fully supply discharge. Then this option reduces the sediment deposition about 5% in the canal under existing sediment inflow rates. While simulations with increased sediment inflow showed that a significant reduction in sediment deposition, about 27%, could be expected. Along with this reduction in sediment deposition the pattern of deposition will also be better as the sediment will be distributed along the canal will not be deposited in one or two head reaches only. Thus it allows a smooth canal operation during the whole operational year without hindrance in the flow deliveries from automatic discharge controllers.

There are some other options of improved canal design for sediment management in the automatically downstream controlled irrigation canals having semi-demand based operation. For example a small increase in the target water level when a raise in water level is observed due to sedimentation, can deliver adequate supplies even under sediment deposition conditions. If flow is reduced due to sedimentation then temporarily the target water level can be increased keeping in view the maximum headloss of the first AVIS/AVIO cross regulator. The decrement of the downstream control gates of silt affected canals can be kept smaller. The larger decrements are more sensitive to changes in water levels, whereas smaller decrements are slow responsive to these variations. If sedimentation takes place in the canal it does not affect immediately the AVIS gate openings due to slow response of the AVIS gate. To maximize the sediment transport capacity of the canal under CBIO the grouping of offtakes for CBIO closures would have to be done all along the canal. In order to discharge the due share of water and sediment to the secondary offtakes in downstream controlled irrigation canals, they would have to be installed close to the AVIS cross regulators at the downstream side. There are minimum water level fluctuations downstream of the cross regulators hence the offtakes will not only draw their fair share of water but they will also draw their fair share of sediments. A lot of mixing of water flow and sediments takes place at the downstream side of the cross regulators due to strong turbulence there. Then it needs some distance for the sediment concentration to the come to the equilibrium concentration. Therefore a significant amount of sediments is discharged into these secondary offtakes. Field measurements have also shown that the offtakes close to cross regulators had about 14% higher sediment withdrawal efficiencies than the other offtakes in the canal.

10.7. Conclusions and way forward

Hydraulic performance of automatically downstream controlled irrigation canals can be improved by improving hydrodynamic stability and reducing the response times, which can be achieved by an optimal combination of PI coefficients and the operation of offtaking canals. Sediment transport is a quite complex process in such canals due to demand or semi-demand based operation. Field measurements and model simulations of flow and sediment transport showed that the canal operation has a significant effect on sediment management. It was found that different operation schemes can reduce sediment deposition even up to 50% in the irrigation canals. Hence the selection of suitable canal operation scheme is equally important as the selection of proper canal design approach for sediment management. The selection of proper operation scheme not only minimizes sediment deposition in the canal but it equally contributes to water savings and environmental sustainability by preventing waterlogging of the irrigated areas. Depending upon the sediment content in the water flow and crop water requirements of the command area, it is possible to erode a fraction of the deposited sediments to minimize the maintenance needs. However, sediment transport in irrigation canals is not only affected by the amount of flow in the canal but other factors like sediment content in the water flow, the operation of the flow regulating structures and canal maintenance all play an important role in defining the sediment transport. The approach of sediment management through improved canal operation works fine for fine sediments in demand based irrigation canals whereas for coarser material additional measures of sediment management would be needed. These measures can be frequent dredging of deposited material or application of a structural approach like sediment traps at appropriate locations, flushing canals, or modifications in canal cross sections. Further studies are needed to see the effectiveness of combining a structural approach with adapted canal operation measures in demand based irrigation canals. While in the coming decades, especially in the emerging countries, many irrigation systems will have to be modernized, such studies will be of major importance with respect to water saving and affordable operation and maintenance.

References

Abbott, M. B., Cunge, J. A. (editors) (1982). Engineering Application of Computational Hydraulics. Vol. 1. Pitman Publishing Ltd. London, United Kingdom.

Ackers, P., White, W. R. (1973). Sediment Transport: new approach and analysis. Journal of Hydraulic Division, American Society of Civil Engineers. Vol. 99, No. 11, New York, USA.

Ackers, P. (1992). Gerard Lacey memorial lecture. Canal and river regime in theory and practice. 1929-92. Proc. Instn. Civ. Engrs. Wat., Marit. and Energy. Technical note No. 619. London, United Kingdom.

Ahmad, M. D., Van Waijjan, E. G., Kupper, M., Visser, S. (1998). Comparison of different tools to assess water distribution in secondary canals with ungated outlets. Research Report No. R-52. International Irrigation Management Institute, Lahore, Pakistan.

Ali, I. (1993). Irrigation and Hydraulic Structures, Theory, design and practice. Institute of Environmental Engineering and Research, NED University of Engineering and Technology, Karachi, Pakistan.

Allen, R. G., Pereira, L. S., Raes, D., Smith, M. (2000). Crop Evapotranspiration. FAO Irrigation and Drainage Paper No. 56.

Alluvial Channel Observation Project (1985). Computer simulation of Tarbela Reservoir sediementation for 1973-1983 period using HEC-6 Model. ACOP-110. Water and Power Development Authority Lahore, Pakistan.

American Society of Civil Engineers. (1972). Sediment Control Methods. C. Control of sediment in canals. Journal of Hydraulic Division. Proceedings of the American Society of Civil Engineers. September, 1972.

Ankum, P. (undated). Lecture notes on Flow Control. UNESCO-IHE, Delft, the Netherlands

Bagnold, R. A. (1956). The flow of Cohesionless grains in fluids. Proceedings, Roy. Soc. London, Phil. Trans. Ser. A., Vol. 249, No. 964.

Bagnold R. A. (1966). An approach to the sediment transport problem from general physics. U. S. Geological Survey Prof. Paper 422.J. Washington, USA.

Barr, D. I. H., Herbertson, J. G. (1966). In discussion of "Sediment Transport Mechanics: Initiation of motion" by the task committee on the preparation of sedimentation manual. Proc. ASCE Journal of Hydraulic Division, Vol. 92, No. HY6. Proc. Paper 4959, Nov. 1966.

Batuca, D. G., Jordaan, J. M. (2000). Silting and desilting of reservoirs. A. A. Balkema, Rotterdam, the Netherlands.

Baume, J. -P., Malaterre, P. -O., Belaude, G., Guennec, B. L. (2005). SIC: A 1D hydrodynamic model for river and irrigation canal modeling and regulation. User manual of SIC model. CEMAGREF, Montpellier, France.

Belaud, G., Baume, J. P. (2002). Maintaining Equity in Surface Irrigation network affected by silt deposition. Journal of Irrigation and Drainage Engineering, vol. 128, No. 5, October, 2002.

Belleudy, P., SOGREAH (2000). Numerical simulation of sediment mixture deposition. Journal of Hydraulic Research 38(6).

Bettess, R., White, W. R., Reeve, C. E. (1988). Width of regime channels. International conference on river regime. Hydraulic research. Wallinglford. United Kingdom.

Bhutta, M. N., Shahid, B. A., Van der Velde, E. J. (1996). Using a hydraulic model to prioritize secondary canal maintenance inputs: results from Pujab, Pakistan. Journal of Irrigation and Drainage Systems, 10: 377-392, 1996.

Blench, T. (1957). Regime behaviour of canals and rivers. Butterworths Scientific Publications, London, England.

Brabben, T. editor (1990). Accuracy of sediment load measurements in rivers and canals. Workshop on sediment measurement and control and deign of irrigation canals. Hydraulic Research, Wallingford, United Kingdom.

Breusers, H. N. (1993). Lecture notes of sediment transport. IHE, Delft, the Netherlands.

Brooks, N. H. (1955). Mechanics of streams with movable beds of fine sand. ASCE, Vol. 123. Paper 2931: 526-594.

Brownlie, W. (1981). Compilation of alluvial channel data: laboratory and field. California Institute of Technology. Report No. KH-R-43B. California, USA.

Carson, M. A., Griffiths, G. A. (1987). Bed load transport in gravel channels. Journal of Hydrology (N. Z.) 26(1).

Celik, I., Rodi, W. (1988). Modelling suspended sediment transport in non equilibrium situation. Journal of Hydraulic Division. ASCE. Vol. 14. New York, USA.

Chang, H. H., Hill, J. C. (1976). Computer modelling of erodible flood channels and deltas, Journal of Hydraulics Division, ASCE, Vol. 102, No. HY10, pp. 132-140.

Chang, H. (1985). Design of stable alluvial canals in a system. Journal of Hydraulic Division, ASCE. New York, USA.

Chow, V. T. (1959). Open Channel Hydraulics. McGraw-Hill, New York.

Colby, B. R., Hembree, D. W. (1955). Computation of total sediment discharge, Niobrara River near Cody, Nebraska. U. S. Geol. Survey, Water Supply Paper 1357.

Colby, B.R. (1957). Relationship of unmeasured sediment discharge to mean velocity. Trans. American Geophysical Union, Vol. 38, No. 5, Oct. 1957.

Cunge, J. A., Holly, F. M., Verwey, A. (1980). Practical Aspects of Computational River Hydraulics, Pitman Publishing Ltd., London, pp. 312-349.

Dehman, E., (1994). Lecture notes on canal design. IHE. Delft, the Netherlands.

Deltares. (2009). User Manual and theoretical background of SOBEK Model. Deltares, Delft, the Netherlands.

Deltares. (2010). Delft3D: Simulation of multidimensional hydrodynamic flows and transport phenomena, including sediment. User Manual, Deltares, Delft, the Netherlands.

Danish Hydraulic Institute (DHI). (2008). Mike 11. Technical reference guide. Danish Hydraulics Institute. Copenhagen, Denmark.

Delft Hydraulics and Ministry of Transport, Public Works and Water Management. (1994). SOBEK: Technical reference guide. Delft Hydraulics. Delft, the Netherlands.

DuBoys, M. P. (1879). Etudes du Regime et I' Action Exercee par les Eaux sur un Lit a Fond de Graviers Indefiniment Affouilable. Annales de Ponts et Chausses, Ser. 5, Vol. 18, pp. 141-195.

Einstein, H. A. (1942). Formulae for the transportation of bed load. Transactions, ASCE, Vol. 107.

Einstein, H. A. (1950). The bed-load function for sediment transportation in open channel flow, Tech. Bulletin 1026. U. S. Department of Agriculture.

Engelund F., Hansen, E. (1967). A monograph on sediment transport in alluvial streams. Teknisk Forlag, Copenhagen, Denmark.

Egiazaroff, I. V. (1950). Coeeficient (f) de la force d'entrainement critique des materieaux par charriage, Proceedings of Ac. Of Sc. Armenian S.S.R. Transl. EDF No. 499.

Fahlbusch, H., Schultz, B., Thatte, C. D. (Editors) (2004). The Indus Basin History of Irrigation, Drainage and Flood Management. ICID, New Delhi, India

Federal Bureau of Statistics. (1999). Population census of Pakistan 1998." Federal Bureau of Statistics, Government of Pakistan, Islamabad, Pakistan.

Galappatti, R. (1983). A depth integrated model for suspended transport. Report No. 83-7, Delft University of Technology, Delft, The Netherlands.

Galappatti, G., Vreugdenhil, C. B. (1985). A depth integrated model for suspended sediment transport. Journal of Hydraulic Research 23(4): 359-357.

Ghumman, A.R., Khan, M. Z., Khan, A. H., Munir, S. (2008). Assessment of operation strategies for logical and optimal use of irrigation water in a downstream control system Irrigation and Drainage 2008.

Gomez, B., Church, M. (1989). "Assessment of bed load sediment transport formulae for gravel bed rivers". Water resources Res. 25(6). 1161-1186.

Graf, W. H. (1971). Hydraulics of Sediment Transport." McGraw-Hill, New York.

Graf, W. H. (2003). "Fluvial Hydraulics – Flow and sediment transport processes in channels of simple geometry." John Wiley & Sons Ltd.

Guo, Q. C., Jin, Y. C. Associate M. ASCE (2002). Modeling Nonuniform suspended Sediment Transport in Alluvial Rivers. Journal of Hydraulic Engineering, Vol. 128, No. 9 Sep 2002.

Haq, I., Abbas, S. T (2006). Sediment of Tarbela and Mangla Reservoirs. Proceedings of 70th annual session of Pakistan Engineering Congress, Paper No. 659. : 23-46.

Harbersack, H. M., Laronne, J. B. (2002), Evaluation and Improvement of Bed Load Discharge Formulae based on Helly-Smith Sampling in an Alpine Gravel Bed River. Journal of Hydraulic Engineering, vol. 128, No. 5, May, 2002.

Holly, F. M., Yang, J. C., Spasojevic, M. (1985). "Numerical simulation of water and sediment movement in multiply-connected networks of Mobile bed channels, IIHR LD Report No. 131. University of Iowa, Iowa City, IA.

Holly, F. M., Preissmann, A. (1977). Accurate calculation of transport in two dimensions. Journal of Hydraulic Division, ASCE, Vol. 103, No. HY11, pp. 1259-1277.

Holly, F. M., Rahuel, J. L. (1990). New numerical/physical framework for mobile modeling Part I: Numerical and physical principles. J. Hydraul. Res. 28(4), 401-416. , 1990

Howard, H. C. (1998). Generalized computer program FLUVIAL-12: Mathematical Model for erodable channels. User Manual. San Diego, California, USA.

Howard, H. C. (2006). Generalized computer program FLUVIAL-12: Mathematical Model for erodable channels. Updated in January 2006. User Manual. San Diego, California, USA.

Hydraulic Research Wallingford. (1992). DORC: user manual. HR Wallingford. Wallingford United Kingdom.

Hydraulic Research Wallingford. (1998). Reservoir Sediment Management, Tarbela Dam. Wallingford, United Kingdom.

IHE. (1998). Duflow User Manual. IHE-Delft. Delft, The Netherlands.

Karim, M. F., Kennedy, J. F. (1982). IALLUVIAL: A computer based flow and sediment routing model for alluvial streams and its application to Missouri River. IIHR Report No. 250. University of Iowa, USA.

Karim, M. F., Kennedy, J. F. (1990). Menu of coupled velocity and sediment-discharge relationship for river. J. Hyd. Eng., 116(8), 987-996.

Kennedy, R. G. (1895). Prevention of silting in irrigation canals. Min. Proc. Inst. Civ. Engrs. Vol. CXIX.

Khan, A. H., Vehmeyer, P. W., Reichert, A. P., Ejaz, M. S., Kalwij, I. M., Lashari, B., Skogerboe, G. V. (1998). Water supply and water balance studies for the Fordwah Eastern Sadiqia (South) Project Area. International Irrigation Management Institute, Lahore, Pakistan.

Knapp, R. T. (1938). Energy balance in stream flows carrying suspended load". Transactions, Amer. Geophy. Union, pp. 501-505.

Kramer, H. (1935). Sand mixtures and sand movement in fluvial models". Transactions, ASCE, vol. 100, pp. 789-878.

Lacey, G. (1929). Stable channels in alluvium. Min. Proc. Inst. Civil Engineers. Vol. 229

Lai, C. T. (1986). Numerical modeling of unsteady open-channel flow. Advances in Hydroscience, Vol. 14. U.S. Geological Survey National Center Reston, Virginia. Pp 225.

Lane, E. W. (1955) The importance of fluvial morphology in hydraulic engineering. Proceedings, ASCE, Vol. 81, No. 745, pp. 1-17.

Lawrence, P. (1990). Canal Design, friction and transport predictor. Workshop on sediment measurement and control and deign of irrigation canals. E.d by T. Brabben. Hydraulic Research, Wallingford, United Kingdom.

Lee, H. Y., Hsieh, H. M. (2003). Numerical simulations of scour and deposition in a channel network. International Journal of Sediment Research. Vol. 18, No. 1, 2003, pp. 32-49.

Lee, H. Y., Hsieh, H. M., Yang, J. C., Yang, C. T. (1997). Quasi two dimensional simulation of scour and deposition in alluvial channels. Journal of Hydraulic Engineering, Vol. 123, No 7.

Liggett, J. A., Cunge, J. A. (1975). Numerical methods of solution of the unsteady flow equation, in K. Mahmood and J. Yevjevich (eds), Unsteady flow in open channels, Vol. 1. Water Resources Publications, Fort Collins.

Linsley, R. K., Franzini, J. B., Freyberg, D. L., Tchobanoglus, G. (1992). "Water Resources Engineering, Fourth ed." McGraw-Hill, New York, USA

Lopes, V. L., Osterkamp, W.R., Bravo-Espinosa, M. (2001). Evaluation of selected bed load equations under transport- and supply-limited Conditions. Proceedings of the Seventh Federal Interagency Sedimentation Conference, 25–29 March, Reno, NV.

Mahmood, K. (1973b). Sediment routing in irrigation canal systems. ASCE, National Water Resources Engineering Meeting. Washington D. C., USA.

Mahmood, K., Chaudhry, A. M., Masood, T. (1978). A guide to hydrologic practice in ACOP –Report No. 48. Alluvial Channel Observation Project (ACOP), WAPDA, Lahore, Pakistan.

Malaterre, P. O., Rogers, D. C., and Schuurmans, J. (1998). Classification of canal control algorithms. Journal of Irrigation and Drainage Engineering 124(1): 23-30

Mantz, P. A. (1977). Incipient transport in fine granins and flakes by fluids-Extended Shield' diagram. J. Hydraulics Div. ASCE 103(HY6): 601-605.

Meadowcroft, I. (1988). The applicability of sediment transport and alluvial friction prediction in irrigation canals. Report technical note OD-TN 34. HR Wallingford. Wallingford, United Kingdom.

Mendez, N. J. V. (1998). Sediment Transport in Irrigation Canals. PhD Thesis. Wageningen Agricultural University, Wageningen, and International Institute for Infrastructure, Hydraulic and Environmental Engineering, Deft, The Netherlands.

Meybeck, M. (1976). Total mineral dissolved transport by world major rivers. Hydrological Science Bulletin 21: 265-284.

Meyer-Peter, E., Muller, R. (1948). Formulae for bed load transport. Proceeding, 3rd Meeting of Intern. Assoc. Hydraulic Res. Stockholm. Pp 39-64.

Milliman, J. D., Meade, R. H. (1983). World-wide delivery of river sediment to the oceans. Journal of Geology 91: 1-21.

Milliman, J. D., Syvitski, J. P. M. (1992). Goemorphic/tectonic control of sediment discharge to the ocean: the importance of small mountainous rivers. Journal of Geology 100: 525-544

Ministry of Water and Power. (2002). "Water Resources of Pakistan." Ministry of Water and Power, Government of Pakistan, Islamabad, Pakistan.

Morris, G. L., Fan, J. (1997). Reservoir sedimentation handbook: Design and management of dams, reservoirs and watersheds for sustainable use. McGraw-Hill, New York, USA.

Munir, S., Kalwij, I. M., Brouwer, M. (1999). Assessment of water distribution at watercourse and minor level of Bahadurwah Minor. Research Report No. R-91. International Irrigation Management Institute, Lahore, Pakistan.

Munir, S., Ahmad, W., Alam, N. (2008). Water saving and environment friendly canal operations in high water allowance canals. Pakistan Journal of Water Resources, Vol. 12(1).

Nakato T. (1990). Test of selected sediment transport formula. Journal of Hydr. Engrg. 116, 362.

Pakistan Institute of Legislative Development and Transparency (PILDAT). (2003). Issues of Water Resources of Pakistan. Publication No. 20-001.

Parker, G. (1990). Surface based bed load transport relation for gravel bed rivers. Journal of Hydraulic Research, Vol. 28(4), 417-436.

Paudel, K. P. (2002). Evaluation of the Sediment Transport Model SETRIC for Irrigation Canals. MSc Thesis. IHE, Delft, The Netherlands.

Paudel, K. P. (2010). Role of sediment in the design and management of irrigation canals. PhD Thesis. UNESCO-IHE, Delft, The Netherlands.

Pongput, K., Alurralde, J. C., Skogerboe, G. V. (1998). Scheduling Model for Crop Based Irrigation Operations. Report No. R-72. International Irrigation Management Institute, Lahore, Pakistan.

Punjab Irrigation and Power Department (2008). Composite schedule rates. Office of the Chief Engineer, Punjab Irrigation and Power Department Lahore, Pakistan http://irrigation.punjab.gov.pk.

Raudkivi, A. J. (1990). Loose boundary hydraulics. 3rd Edition. Pergamon Press Canada.

Reid, I., Laronne, J.B. (1995). Bed load sediment transport in an ephemeral stream and a comparison with seasonal and perennial counterparts. Water Resources Research 31(3): 773-781.

Reid, I., Powell, D.M., Laronne, J.B. (1996). Prediction of bed load transport by desert flash-floods. Journal Hydraulic Engineering 122: 170-173.

Rodellar, J., Gomez, M., Bonet, L. (1993). Control method for on demand operation of open channel flow. Journal of irrigation and Drainage Engineering, ASCE 119(2): 225-221.

Rouse, H. (1937). Modern conceptions of the mechanics of turbulence. Transactions ASCE, Vol. 102.

Rouse, H. (1939). An analysis of sediment transportation in light of Fluid Turbulence. SCS-TP-25. Washington, DC: Soil Conservation Service, U. S. Department of Agriculture

Rubey, W. W. (1933). Equilibrium-conditions in debris-laden streams. Transactions, Amer. Geophy. Union.

Sawadogo, S., Achaibou, A. K., Aguilar-Martin, J., Mora-Camino, F. (1992b). Intelligent control of large water distribution systems: a two level approach. SICICI 92, Singapour, Proc. 1085-1089.

Schoklitsch, A. (1930). Hanbuch des Wasserbaues. Springer, Vienna. English translation (1937) by S. Shulits.

Seborg, D. E., Edgar, T. F., Mellichamp, D. A. (1989). Process dynamics and control. New York, John Wiley.

Shames, I. N. (1962). Mechanics of Fluids. McGraw Hill, New York

Shand, M. J. (1971). Automatic downstream control systems for irrigation canals. PhD Thesis, University of California, Berkeley, California.

Shen, H. W., Hung, C. S. (1971). An Engineering approach to total bed material load by regression analysis. Proceedings, Sedimentation Sym. Berkeley.

Sheng, Y. P. L., Butler, H. (1982). Modeling coastal currents and sediment transport. Proc. 18th Int. Conf. Coastal Eng. 1127.

Shields, A. (1936). Anwendung der Aehnlihk eistmechanik und Turbulenz forschung auf die Geschiebewe gung. Mitteilung Preussischen versuch anstalt wasser, Erd, Schi ffbav, belin, No. 26 Germany.

Shou-Shan, F. (1989). An overview of computer stream sedimentation models. Sediment transport modeling. Ed. By S. S. Y. Wang. ASCE. New York, USA.

Shulits, S., Hill, R. D. Jr. (1968). Bed load formulae. Bulletin, Department of Civil Engineering, Hydraulics Laboratory, Pennsylvania State University, Park PA.

Simons, D. B., Albertson, M. L. (1960). Uniform water conveyance channels in alluvial material. Journal of Hydraulic Division, ASCE, 86(HY5), 33-71.

Simons D. B., Senturk, F. (1976). Sediment Transport Technology. Water Resources publications, Colorodo USA.

Simons, D. B., and Senturk, F. (1977). Sediment Transport Technology. Water Resources Publications, Fort Collins, Co.

Simons, D. B., Li, R-M., Cotton, G. K. (1984). Mathematical model for estimating scour through bridge crossings. Transportation Research Record 2(950).

Simons, D. B., and Senurk, F. (1992). Sediment Transport Technology. Water Resources Publications, Fort Collins, Co.

Spasojevic, M., Holly, F. M. (1988). Numerical simulation of two-dimensional deposition and erosion patterns in alluvial water bodies. IIHR Report No. 149, the University of Iowa, Lowa City, Iowa, U.S.A.

Surface Water Hydrology Project (2008). Project report No. 65. Water and Power Development Authority Lahore, Pakistan.

Thomas, W. A., Prasuhn, A. L. (1977). Mathematical modeling of scour and deposition. J. Hydr. Div. ASCE, 103(8), 851-863.

Tison, L. J. (1953). Recherches sur la Tension Limite d' Entrainment des Materiaux Constittifs du Lit. IAHR, 5th Congress, Minneapolis.

Timilsina, P. (2005). One-Dimensional Convection Diffusion Model for Non-equilibrium Sediment Transport in Irrigation Canals. MSc Thesis, UNESCO-IHE, Delft, The Netherlands.

Toffaletti, F. B. (1969). Definitive computations of sand discharge in rivers. J. of the Hydraulic Div. ASCE, Vol. 95, No. HY1, January, pp. 225-246.

U.S. Bureau of Reclamation. (2010). SRH 1D: Sedimentation and river hydraulics models. User Manual. United States Department of Interior, USA.

Vanoni, V. A., Brooks, N. H., Kennedy, J. F. (1961). Lecture notes on sediment transportation and channel stability. Report KH-R-1, W. M. Kech Lab. of Hydraulics and Water Resources, California Institute of Technology, Pasadena, California.

Vanoni, V. A. (Ed.). (1966). Journal of Hydraulics Division, ASCE, vol. 92. No. HY6, Proceedings paper 4959.

Van Rijn, L. C. (1984a). Sediment Transport Part I: Bed Load Transport. Journal of Hydraulic Division, ASCE. Vol. 110, No. 10, New York, USA.

Van Rijn, L. C. (1984b). Sediment Transport Part II: Suspended Load Transport. Journal of Hydraulic Division, ASCE. Vol. 110, No. 11, New York, USA.

Van Rijn, L. C. (1984c). Sediment Transport Part III: Bed forms and alluvial roughness. J. Hydraul. Eng., 110(2), 1733-1754.

Van Rijn, L. C. (1986). Mathematical modeling of suspended sediment in nonuniform flows. J. Hydr. Engrg, ASCE, 112(6), 433-455

Van Rijn, L. C., Meijer, K. (1988). Three dimensional mathematical modelling of suspended sediment transport in currents and waves. Delft Hydraulics, Delft, the Netherlands

Vreugdenhil, C. B. (1989). Computational Hydraulics. Springer-Verlag. Berlin, Germany.

Vreugdenhil, C. B., Vries, M. de. (1967). Computations on Non-steady Bedload Transport by a Pseudo-viscosity Method. Hydraulic Laboratory Delft, the Netherlands.

Vreugdenhil, C. B., Wijbenga, J. H. A. (1982). Computation of flow patterns in rivers. Journal of Hydraulic Division, ASCE, 108(11): 1296-1310.

Wang, S. S. Y., Adeff, S.E. (1986). Three-dimensional modelling of river sedimentation processes. Proc. of the 3rd Int. Symp. on River Sedimentation, The Univ. of Mississipi, USA.

Water and Power Development Authority (WAPDA). (1985). ACOP-Third Bi-national symposium on Mechanics of Alluvial Channels held in Lahore, March 3-6, 1985

White, C. H. (1940). The equilibrium of grains on the bed of a stream. Proc. Roy. Soc. London, Vol. 174A.

Woo, H. (1998a). Performance test of some selected transport formulae. Journal of Hydraulic Division, ASCE, New York, USA.

World Bank. (2007). Pakistan promoting rural growth and poverty reduction. Sustainable and Development Unit, World Bank.

Yang, C. T. (1972). Unit stream power and sediment transport. J. Hydr. Div., Proc. ASCE 98 (HY10): 1805-1856.

Yang, C. T. (1973). Incipient Motion and sediment transport. Journal of Hydraulic Division, ASCE. Vol. 99. No. 110. New York, USA.

Yang, C. T. (1984). Unit Stream Power equation for gravel. J. Hydr. Div. Proc. ASCE 110(HY12).

Yang, C. T. (1996). Sediment transport, theory and practice. McGraw Hill, New York.

Zanke, U. (1990). Der Beginn der Sedimentbewegung als Wahrscheinlichkeitsproblem (The Beginning of Sediment Motion as a Probability Problem, in German). Wasser and Boden, Heft 1.

Zanke, U.C.E. (2003). On the influence of turbulence on the initiation of sediment motion. International Journal of Sediment Research, vol. 18, No. 1, 2003, pp. 17-31.

List of symbols

Symbol	Description	Unit
A	Cross-sectional area	m^2
A_b	Solid area of channel bed	m^2
B	Bed width	m
C	Chezy coefficient	$m^{1/2}/s$
C	Sediment concentration	m^3/m^3, kg/m^3
C_a	Reference sediment concentration	m^3/m^3
$\bar{c}$	required concentration at a distance x from the origin	ppm
$\bar{c}_e$	equilibrium concentration at x from the origin	ppm
$\bar{c}_o$	actual concentration at the origin	ppm
d_{50}	Median diameter	
D_*	dimensionless grain parameter	
e	Error	m
e_b	Bed load transport efficiency = bed load work rate/stream power	
e_s	Suspension efficiency = suspension work rate/stream power	
Fr	Froude Number	
g	Acceleration due to gravity	m/s^2
H	Total head	m
i	Subscript for length	
j	Superscript for time	
K	Channel's conveyance	
K_D	Diffusion coefficient	m^2/s
K_p	Proportional gain	
k_{se}	Equivalent roughness	m
k_s'	roughness due to grain	m
k_s	roughness due to form developed in the bed	m
L	Length of canal reach	m
L_A	adaptation length	(m)
m	Side slope	V : 1H
n	Manning's roughness coefficient	$s/m^{1/3}$
N	exponent of velocity in sediment transport equation	
P	Wetted perimeter	m
p	Probability for a particle to be eroded at any one spot	
Q	Water discharge	m^3/s

Symbol	Description	Unit
Q_s^*	equilibrium sediment transport capacity	(m^3/s)
Q_s	actual sediment discharge in the canal	(m^3/s)
Q_sL	lateral sediment discharge (input or output)	m^3/s
q	Lateral inflow outflow	m^2/s
q_s	Sediment discharge per unit width	m^2/s
R	Hydraulic radius	m
r	Float radius	m
S_f	Energy slope	m/m
S_o	Bed slope	m/m
T_d	Derivative control action	
T_i	Integral action time	seconds
T_u	Travel time upward	seconds
T_w	Travel time downward	seconds
u_*	Shear velocity	m/s
E_p	Available power	
t	Time	seconds
u	Longitudinal velocity component, point velocity	m/s
U	Mean flow velocity	m/s
V_{dyn}	Positive storage wedge	m^3
W	Water surface width at full supply discharge	m
$\bar{w}$	Vertical velocity component	m/s
w_f	Fall velocity	m/s
x	Longitudinal distance	m
y	Water depth	m
z	Water level	m
α	Calibration parameter for adaptation length	
β	Calibration parameter for sediment transport predictor	
Δx	Length Step	m
Δt	Time step	seconds
Δy	Maximum change is water level in AVIS/AVIO gates	m
φ	sediment exchange rate with the bed	
ε	Decrement	m
θ	Weighting coefficient	
ψ	Shear intensity parameter	
ε_s	Sediment diffusion coefficient	
ε	Turbulence eddy viscosity	

Symbol	Description	Unit
ρ	Density of water	kg/m^3
σ	Courant number	
τ_o	Bed shear stress	N/m^2
κ	von Karman's constant	

Acronyms

Absc.	Abscissa
ACOP	Alluvial Channel Observation Project
AOSM	Adjustable Orifice Semi Module
APM	Adjustable proportional module
BCM	Billion Cubic Meter
CBIO	Crop Based Irrigation Operations
CRM	Coefficient of residual mass
CWR	Crop Water Requirement
Disty	Distributary
DPR	Delivery Performance Ratio
EF	Efficiency
ETR	Equi-Transit Rate
EWI	Equi-Width Increment
FANA	Federally Administered Northern Area
FATA	Federally Administered Tribal Area
FB	Feed Back
FDM	Finite Difference Method
FEM	Finite Element Method
FF	Feed Forward
FSE	Full Supply Extended
GK	Ghazikot
GO	Gandaf outlet
GVF	Gradually varied flow
HMI	Human Machine Interface
HR	Hydraulic Research
IBIS	Indus Basin Irrigation System
ISRIP	International Sedimentation Research Institute
MAE	Maximum Absolute Error
MCM	Million Cubic Meter
ME	Maximum error
MIMO	Multiple Input Multiple Output
MISO	Multiple Input Single Output
MMT	Million Metric Tonne
MOC	Method of Characteristics
MSL	Mean Sea Level
NN	Neknam
O&M	Operation and Maintenance
PHLC	Pehure High Level Canal
PID	Provincial Irrigation Department
PSD	Particle Size Distribution
RMSE	Root Mean Square Error
SCADA	Supervisory Control and Data Acquisition

SETRIC	SEdiment TRansport In Irrigation Canals
SIC	Simulation of Irrigation Canals
SISO	Single Input Single Output
USC	Upper Swat Canal
USGS	United Stated Geological Survey
VAT	Visual Accumulation Tube
WAPDA	Water and Power Development Authority
WL	Water Level
XR	Cross Regulator
YB	Yaqubi
YH	Yar Hussain

Appendices

A. Sediment transport relationships under equilibrium conditions

A number of widely used equilibrium sediment transport formulae have been discussed in the following sections.

Bagnold (1966) formula. Bagnold formula is based on energy concepts. Bagnold assumed that the power necessary to transport sediment is proportional to the loss of potential energy of the flow per unit time, which Bagnold called "available power". The power per unit area can be written as:

$$E_p = \rho g y S_f U = \tau_o U \tag{A.1}$$

Considering bed load and suspended load transport separately, Bagnold obtained the following formula:

$$\rho g \frac{q_s}{\tau_o U} = \frac{e_b}{f} + e_s (1 - e_b) \frac{u_s}{w_f} \tag{A.2}$$

Assuming $u_s = U$ and proposing numerical values to make his formula practical, he finally obtained:

$$q_s = \frac{\tau_o U}{\rho g} \left[\frac{e_b}{f} + 0.01 \frac{U}{w} \right] \tag{A.3}$$

Where
ρ	= density of fluid (kg/m^3)
g	= acceleration due to gravity (m/s^2)
y	= water depth (m)
S_f	= energy slope (m/m)
U	= mean flow velocity (m/s)
E_p	= Available power
q_s	= sediment discharge per unit width (m^2/s)
τ_o	= bed shear stress (N/m^2)
u_s	= velocity of the sediments (m/s)
e_b	= efficiency of the flow for bed load
e_s	= efficiency for suspended load
w_f	= fall velocity (m/s)
f	= tanψ a friction factor on the bed

For sand finer than 0.50 mm and Y (Shields' dimensionless shear stress) > 1, Bagnold gives the value of 0.17 for e_b/f and special graphs to assess it, see Graf (1971).

Engelund and Hansen formula (1967). The formula of Engelund and Hansen is based on the energy approach. They established a relationship between transport and mobility parameter. The Engelund and Hansen function for the total sediment transport is calculated by:

Dimensionless transport parameter ϕ is give by:

$$\phi = \frac{q_s}{\sqrt{(s-1)gd_{50}^3}} \qquad (A.4)$$

Dimensionless mobility parameter θ is given by:

$$\theta = \frac{u_*^2}{(s-1)gd_{50}} \qquad (A.5)$$

The relationship between these parameters is expressed by:

$$\phi = \frac{0.1\theta^{2.5}C^2}{2g} \qquad (A.6)$$

And the total sediment transport is expressed by:

$$q_s = \frac{0.05U^5}{(s-1)^2 g^{0.5} d_{50} C^3} \qquad (A.7)$$

Where
q_s	= total sediment transport (m²/s)
θ	= dimensionless mobility parameter
ϕ	= dimensionless transport parameter
U	= mean velocity (m/s)
q_s	= sediment discharge per unit width (m²/s)
C	= Chezy coefficient (m$^{1/2}$/s)
s	= relative density
u_*	= shear velocity (m/s)
d_{50}	= mean diameter (m)
g	= acceleration due to gravity (m/s²)

Ackers and White (1973). The Ackers and White formula is based on Bagnold's stream power concepts. They developed a general sediment-discharge function in terms of three dimensionless groups: D_* (size), F_{gr} (mobility), G_{gr} (discharge).

Dimensionless grain diameter D_* is given by:

$$D_* = D_{35}\left[g(s-1)/v^2\right]^{1/3}$$
(A.8)

Dimensionless mobility number F_{gr} is given by:

$$F_{gr} = \frac{u_*^n}{\sqrt{gd_{35}(s-1)}}\left(U \middle/ \log\left(\frac{10y}{d_{35}}\right)\sqrt{32}\right)^{1-n}$$
(A.9)

with $n = 1.00 - 0.56\ logD_*$

The dimensionless transport parameter, G_{gr}, is based on the stream power concept. The general equation for this parameter is:

$$G_{gr} = c\left(\frac{F_{gr}}{A} - 1\right)^m$$
(A.10)

With $A = 0.23/\sqrt{D_*} + 0.014$ and $m = 9.66/D_* + 1.334$

$$\log c = 2.86 \log D_* - (\log D_*)^2 - 3.53$$
(A.11)

The Ackers and White function to determine the total sediment transport is given as:

$$q_s = G_{gr}sd_{35}\left(\frac{U}{u_*}\right)^n$$
(A.12)

Where

D_*	= dimensionless grain parameter
s	= relative density
g	= acceleration due to gravity (m/s^2)
d_{35}	= representative particle diameter (m)
y	= water depth (m)
v	= kinematic viscosity (m^2/s)
F_{gr}	= dimensionless mobility parameter
A	= Value of F_{gr} at nominal, initial movement
G_{gr}	= dimensionless transport parameter
c	= coefficient in transport parameter G_{gr}
m	= exponent in the transport parameter G_{gr}
n	= exponent in the dimensionless mobility parameter F_{gr}
u_*	= shear velocity (m/s)
U	= mean flow velocity (m/s)
q_s	= total sediment transport per unit width (m^2/s)

Yang's (1973). Yang's formula is also based on the concept of unit stream power. The unit stream power is defined as the power available per unit weight of fluid to transport sediment and is equal to the product of mean flow velocity and energy slope, *US*. The formula developed on the basis of multiple regression analysis of 463 set of laboratory data for sand transport in terms dimensionless variables gave the following relationship for total sediment discharge:

$$\log C_t = 5.44 - 0.29 \log \frac{w_f d_{50}}{v} - 0.46 \log \frac{u_*}{w_f}$$

$$+ \left(1.80 - 0.409 \log \frac{w_f d_{50}}{v} - 0.314 \log \frac{u_*}{w_f} \right) \qquad (A.13)$$

$$* \log \left(\frac{US_f}{w_f} - \frac{U_{cr} S_f}{w_f} \right)$$

$$\frac{U_{cr}}{w_f} = \frac{2.5}{\log \frac{u_* d_{50}}{v} - 0.06} + 0.06 \quad for \quad 1.2 < \frac{u_* D_{50}}{v} \qquad (A.14)$$

$$\frac{U_{cr}}{w_f} = 2.05 \qquad for \quad 70 > \frac{u_* D_{50}}{v} \qquad (A.15)$$

The total load transport is calculated by:

$$q_s = 0.001 c_t U y \qquad (A.16)$$

Where

$\log C_t$	= total sand concentration by weight in ppm = $10^6 \gamma_s q_t / \gamma q$
u_*	= shear velocity (m/s)
U_{cr}	= flow velocity (m/s) at incipient motion
U	= flow velocity (m/s)
q_s	= total load transport (kg/s/m)
w_f	= fall velocity (m/s)
C_t	= total sediment transport in ppm by mass
d_{50}	= particle size (m)
v	= kinematic viscosity (m²/s)
u_*	= shear velocity (m/s)
y	= water depth (m)
S_f	= energy slope (m/m)

Brownlie (1981). Brownlie formula to compute the sediment transport capacity is based on dimensional analysis and calibration of a wide range of field and laboratory data, where uniform conditions prevailed. The transport (in ppm by weight) is calculated as:

$$q_s = 728 \, c_f \, (F_g - F_{g_{cr}})^{1.98} \, S_f^{0.66} \left(\frac{R}{d_{50}} \right)^{-0.33}$$ (A.17)

Grain Froude number, F_g, is:

$$F_g = \frac{U}{\left[(s-1) g d_{50} \right]^{0.5}}$$ (A.18)

Critical grain Froude number, F_{gcr}, is:

$$F_{g_{cr}} = 4.60 \, \tau_{*_o}^{0.53} \, S_f^{-0.14} \, \sigma_s^{-0.17}$$ (A.19)

Critical dimensionless shear stress, τ_{*o} is:

$$\tau_{*_o} = 0.22 \, Y + 0.06(10)^{-7.7Y}$$ (A.20)

The Y value is given as:

$$Y = (\sqrt{s-1} \, R_g)^{-0.6}$$ (A.21)

Grain Reynolds number, R_g, is:

$$R_g = \frac{(g d_{50}^3)^{0.5}}{31620 \, v}$$ (A.22)

Where

c_f	= coefficient for the transport rate ($c_f = 1$ for laboratory and $c_f = 1.268$ for field conditions)
F_g	= grain Froude number
F_{gcr}	= critical grain Froude number
τ_{*o}	= critical dimensionless shear stress
σ_s	= geometric standard deviation
g	= acceleration due to gravity (m/s^2)
S_f	= energy slope (m/m)
d_{50}	= median diameter (mm)
s	= relative density
R_g	= grain Reynolds number
R	= hydraulic radius (m)
v	= kinematic viscosity (m^2/s)

Van Rijn Method (1984a and 1984b). The total sediment load transport by the van Rijn method can be computed by summation of the bed and suspended load transport.

The van Rijn method presents the computation of the bed load transport qb as the product of the saltation height, the particle velocity and the bed load concentration. It is assumed that the motion of the bed particles is dominated by gravity forces.

Bed load transport rate is calculated as:

$$q_b = u_b \delta_b c_b \qquad\qquad (A.23)$$

$$u_b = 1.5T^{0.6}((s-1)gd_{50})^{0.5}$$

$$\delta_b = 0.3D_*^{0.7}T^{0.5}d_{50}$$

$$c_b = 0.18c_oT/D_*$$

with:

$$T = \frac{(u_*')^2 - (u_{*cr})^2}{(u_{*cr})^2}$$

$$u_*' = g^{0.5}U/c$$

$$D_* = ((s-1)g/v^2)^{1/3}d_{50}$$

The bed load transport rate can be expressed as:

$$q_b = 0.053(s-1)^{0.5}g^{0.5}d_{50}^{1.5}D_*^{-0.3}T^{2.1} \qquad\qquad (A.24)$$

Where

q_b	= bed load transport rate (m^3/s)
T	= bed shear parameter
D_*	= Particle parameter
s	= relative density
g	= acceleration due to gravity (m/s^2)
u_b	= particle velocity (m/s)
c_b	= bed load concentration (ppm)
c_o	= maximum volumetric concentration = 0.65
D_*	= particle diameter
u_*'	= shear velocity (m/s)
C'	= Chezy coefficient related to grains (m$^{1/2}$/s) = 18 log (12y/3d$_{90}$)
δ_b	= saltation height (m)
d_{50}	= mean diameter (m)
g	= gravity acceleration (m/s^2)

The suspended load transport q_{sus} is calculated by:

$$q_{sus} = FUyc_a \tag{A.25}$$

Where
  F = shape factor
  U = mean flow velocity (m/s)
  y = water depth (m)
  c_a = reference concentration (ppm)

$$F = \frac{(a/y)^{z'} - (a/y)^{1.2}}{(1 - a/y)^{z'}(1.2 - z')} \tag{A.26}$$

Suspension parameter:

$$z = \frac{w_f}{\beta \kappa u_*}$$

Modified suspension parameter: $z' = z + \psi$

Reference concentration:

$$c_a = \frac{0.015 d_{50} T^{1.5}}{a D_*^{0.3}}$$

Reference level: $a = 0.5\Delta$ or $a = k_s$ with $a_{min} = 0.01y$

Representative size of suspended sediment d_s:

$$d_s = [1 + 0.011(\sigma_s - 1)(T - 25)]d_{50} \tag{A.27}$$

β-factor:
$$\beta = 1 + 2(\frac{w_f}{u_*})^2 \tag{A.28}$$

ψ-factor:

$$\psi = 2.5(\frac{w_f}{u_*})^{0.8}(\frac{c_a}{c_o})^{0.4} \tag{A.29}$$

Where

F	= shape factor
U	= mean flow velocity (m/s)
u_*	= shear velocity (m/s)
u^*_{cr}	= critical shear velocity (m/s)
c_a	= reference concentration
y	= water depth (m)
D_*	= dimensionless particle parameter
a	= reference level (m)
z	= suspension number
z'	= modified suspension number
β	= ratio of sediment and fluid mixing coefficient
ψ	= stratification correction
k	= constant of von Karman
σ_s	= geometric standard deviation
d_{50}	= median diameter (mm)
d_s	= representative particle size of suspended sediment (m)
w_f	= fall velocity of representative particle size (m/s)
T	= transport stage parameter
Δ	= bed form height (m)
k_s	= equivalent roughness height (m)

B.   The Saint-Venant equations and their solution

In open channel flow the Saint-Venant equations fully describe the hydrodynamics of the unsteady flow. These equations are based on the basic laws of conservation of mass, the continuity equation (Eq. B.1), and the conservation of momentum, the momentum equation (Eq. B.2). These equations are presented with y(x, t) and Q(x, t) as dependent variables as follows (Cunge et al., 1980):

-     Continuity equation:

$$\frac{\partial y}{\partial t} + \frac{1}{b}\frac{\partial Q}{\partial x} = 0 \qquad\qquad (B.1)$$

-     Momentum equation:

$$\frac{\partial Q}{\partial t} + \frac{\partial}{\partial x}(\frac{Q^2}{A}) + gA\frac{\partial y}{\partial x} - gAS_o + gA\frac{Q|Q|}{K^2} = 0 \qquad\qquad (B.2)$$

Where
y	= water depth (m)
Q	= discharge (m^3/s)
x	= distance (m)
t	= time (s)
b	= top width (m)
A	= cross section area (m)
K	= the conveyance and becomes $S_f K^2 = Q|Q|$
V	= average flow velocity (m/s)
S_o	= Bed slope (m/m)
S_f	= energy slope (m/m)

Solution techniques

The Saint-Venant equations are the non-linear hyperbolic partial differential equations. They do not have any known analytical solution. Hence they are solved numerically by number of different methods. The three known methods of solution are the Method of Characteristics (MOC), Finite Element Method (FEM) and the Finite Difference Method (FDM). The FEM and MOC are rarely used for one dimensional (1-D) unsteady flow problems, whereas the FDM is mostly used for 1-D problems (Cunge et al., 1980).

Finite difference method

In finite difference method the state of the flow describing continuous functions are replaced by functions defined on a finite number of grid points within the considered domain. The derivatives are then replaced by divided differences. Then, the differential equations are replaced by algebraic finite difference relationships. The ways in which derivatives are expressed by discrete functions are termed as finite difference schemes.

The computational grid, Figure B.1, is a finite set of points containing the same domain in the (x, t) plane as the continuous argument functions. This set is the domain of definition of the discrete-argument functions which is called grid functions.

The finite difference schemes used for unsteady state flow modelling can be grouped into different classes, based upon their characteristics. Cunge et al, 1980 described that the main distinction among different schemes in a class is related to the way in which physical coefficients in the flow equations, like $A(y)$, $B(y)$, $S_f(y, Q)$ in Equation B.1 and B.2 are discretized. The variability of coefficients makes the equations non-linear due to their dependence on the flow variables. The finite difference analogues of the basic differential equations become then non-linear algebraic systems.

A number of different finite difference schemes are used for unsteady state flow modelling, which are broadly categorized as explicit and implicit schemes. The spatial partial derivatives replaced in terms of variable at known time level are referred as explicit finite differences, whereas those in terms of variables at the unknown time level are called implicit finite difference (Chaudhry, 1994). The typical finite difference schemes used for unsteady state flow modelling are the Lax Diffusive scheme, Mac-Cormack scheme, Leap-Frog scheme and the Gabutti scheme (Chaudhry, 1994).Whereas the commonly used implicit finite difference schemes are the Abbot-Ionesco scheme, Delft Hydraulics Laboratory scheme and the Preissmann scheme. The Preissmann scheme is the most widely used in unsteady state flow modelling, which is unconditionally stable scheme and gives accurate results.

Preissmann scheme

Cunge et al, (1980) presented the application of Preissmann scheme to the derivatives in Equations B.1 and B.2 with $\psi = 0.5$ as follows:

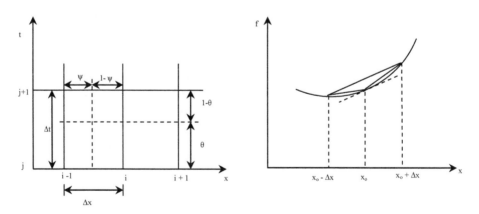

Figure B.1: Computational grid for Preissmann scheme and approximations

$$\frac{\partial y}{\partial t} \approx \frac{y_{i+1}^{j+1} - y_{i+1}^{j}}{2\Delta t} + \frac{y_{i}^{j+1} - y_{i}^{j}}{2\Delta t} \tag{B.3}$$

$$\frac{\partial Q}{\partial t} \approx \frac{Q_{i+1}^{j+1} - Q_{i+1}^{j}}{2\Delta t} + \frac{Q_{i}^{j+1} - Q_{i}^{j}}{2\Delta t} \tag{B.4}$$

$$\frac{\partial y}{\partial x} \approx \theta \frac{y_{i+1}^{j+1} - y_{i}^{j+1}}{\Delta x} + (1-\theta)\frac{y_{i+1}^{j} - y_{i}^{j}}{\Delta x} \tag{B.5}$$

$$\frac{\partial Q}{\partial x} \approx \theta \frac{Q_{i+1}^{j+1} - Q_{i}^{j+1}}{\Delta x} + (1-\theta)\frac{Q_{i+1}^{j} - Q_{i}^{j}}{\Delta x} \tag{B.6}$$

$$\frac{\partial}{\partial x}\left(\frac{Q^2}{A}\right) \approx \frac{\theta}{\Delta x}\left[\frac{(Q_{i+1}^{j+1})^2}{A_{i+1}^{j+1}} - \frac{(Q_{i}^{j+1})^2}{A_{j}^{n+1}}\right]$$
$$+\frac{(1-\theta)}{\Delta x}\left[\frac{(Q_{i+1}^{j})^2}{A_{i+1}^{j}} - \frac{(Q_{i}^{j})^2}{A_{i}^{j}}\right] \tag{B.7}$$

The space interval in the above equations is $\Delta x = x_{i+1} - x_i$

The coefficients in the Equations B.1 and B.2, when represented according to Preissmann formulation yield together with the derivatives expressed by Equations B.2 and B.4 to the following two algebraic equations B.9 and B.10:

$$f(x,t) \approx \frac{\theta}{2}(f_{i+1}^{j+1} + f_{i}^{j+1}) + \frac{1-\theta}{2}(f_{i+1}^{j} + f_{i}^{j}) \tag{B.8}$$

$$\frac{1}{2}\left(\frac{y_{i+1}^{j+1} - y_{i+1}^{j}}{\Delta t} + \frac{y_{i}^{j+1} - y_{i}^{j}}{\Delta t}\right)$$
$$+\frac{2}{\Delta x}\frac{\left[\theta(Q_{i+1}^{j+1} - Q_{i}^{j+1}) + (1-\theta)(Q_{i+1}^{j} - Q_{i}^{j})\right]}{\theta(b_{i}^{j+1} + b_{i+1}^{j+1}) + (1-\theta)(b_{i}^{j} + b_{i+1}^{j})} = 0 \tag{B.9}$$

$$\frac{1}{2}\left(\frac{Q_{i+1}^{j+1}-Q_{i+1}^{j}}{\Delta t}+\frac{Q_{i}^{j+1}-Q_{i}^{j}}{\Delta t}\right)+\frac{\theta}{\Delta x}\left[\left(\frac{Q^{2}}{A}\right)_{i+1}^{j+1}-\left(\frac{Q^{2}}{A}\right)_{i}^{j+1}\right]$$

$$+\frac{(1-\theta)}{\Delta x}\left[\left(\frac{Q^{2}}{A}\right)_{i+1}^{j}-\left(\frac{Q^{2}}{A}\right)_{i}^{j}\right]+g\left[\theta\frac{A_{i+1}^{j+1}+A_{i}^{j+1}}{2}+(1-\theta)\frac{A_{i+1}^{j}+A_{i}^{j}}{2}\right]$$

$$\left(\begin{array}{l}\left[\theta\dfrac{y_{i+1}^{j+1}-y_{i}^{j+1}}{\Delta x}+(1-\theta)\dfrac{y_{i+1}^{j}-y_{i}^{j}}{\Delta x}\right]+\\[4mm]\left[\theta\dfrac{Q_{i+1}^{j+1}\mid Q_{i+1}^{j+1}\mid+Q_{i}^{j+1}\mid Q_{i}^{j+1}\mid}{2}+(1-\theta)\dfrac{Q_{i+1}^{j}\mid Q_{i+1}^{j}\mid+Q_{i}^{j}\mid Q_{i}^{j}\mid}{2}\right]\\[4mm]\left[\theta\dfrac{(K_{i+1}^{j+1})^{2}-(K_{i}^{j+1})^{2}}{2}+(1-\theta)\dfrac{(K_{i+1}^{j})^{2}+(K_{i}^{j})^{2}}{2}\right]^{-1}\end{array}\right.=0\,;0.5\leq\theta\leq1.0$$

$$(B.10)$$

Here again two non-linear algebraic equations appear in terms of $y_{i}^{j+1}, Q_{i}^{j+1}, y_{i}^{j}$ and Q_{i}^{j}. The way the coefficients leading to Equations B.9 and B.10 are expressed also influence the way in which the non-linear system is solved. If it is supposed that the all functions f(y, Q) in the discretized non-linear algebraic equations are known at time level nΔt and are differentiable with respect to y and Q. Then these functions can be evaluated a the time level (n+1)Δt by a Taylor series expansion of the form:

$$f_{i}^{j+1}=f_{i}^{j}+\Delta f=f_{i}^{j}+\frac{\partial f_{i}}{\partial y}\Delta y_{i}+\frac{\partial f_{i}}{\partial Q}\Delta Q_{i}+\frac{\partial^{2}f}{\partial y^{2}}\frac{\Delta y_{i}^{2}}{2}+....\qquad(B.11)$$

Where Δy, ΔQ are y and Q increments during the time step Δt. The substitution of this expansion into the finite difference approximations leads to a system of two non-linear algebraic equations in terms of Δy_i, ΔQ_i, Δy_{i+1}, ΔQ_{i+1} for every pair of point (i, i+1). For N computational points there will be a system of 2N-2 such equations for 2N unknowns. With the addition of two boundary conditions the system can be solved by Newton iteration method, and for that purpose it is first linearized in terms of unknowns Δy_i, ΔQ_i, i = 1, 2, 3, ..., N as described by Ligget and Cunge (1975). For a pair of adjacent points (i, i+1) the linearized equations can be written as follows:

$$A1\Delta y_{i+1}+B1\Delta Q_{i+1}+C1\Delta y_{i}+D1\Delta Q_{i}+G1=0$$
$$A2\Delta y_{i+1}+B2\Delta Q_{i+1}+C2\Delta y_{i}+D2\Delta Q_{i}+G2=0\qquad(B.12)$$

Using the known values of y_{i}^{j}, Q_{i}^{j} the coefficients A1, B1, C1, ..., G2' can be computed, and the linearized system of Equations B.12 written for N point can be solved using tri-diagonal matrices.

The coefficients A_i, B_i, C_i, ..., G_i can be computed at the time step t^j, representing the initial conditions, where the values of h_i^j, Q_i^j, h_{i+}^j and Q_{i+}^j are known. These two linear algebraic equations can be used for the adjacent points (i, i+1). However, they are not yet sufficient to find the four unknowns, Δh_i, ΔQ_i, Δh_{i+1}, ΔQ_{i+1}. With addition of two boundary conditions, upstream and downstream, the system can be solved. Then the double sweep method is used to obtain the solution of the system of the equations (Liggett and Cunge, 1975).

C.   One dimensional convection diffusion equation and its solution in unsteady state flow conditions

The mass balance equation for a conservative matter in water with concentration c can be written as in x (longitudinal) direction:

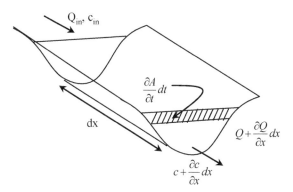

Figure C.1. Control element for the derivation of convection-diffusion equation

$$\frac{\partial}{\partial}(Ac) + \frac{\partial}{\partial x}(Qc) = 0 \qquad (C.1)$$

By differentiating out the terms of the Equation C.1 and substituting the continuity equation of flow (Eq. B.1) it turns to the following equation:

$$\frac{\partial c}{\partial t} + U\frac{\partial c}{\partial t} = 0 \qquad (C.2)$$

Further, the advection through the element boundary is increased by the contribution of molecular and turbulent exchange processes, which can be described by Fick's Law of diffusion as:

$$T_{dispersion} = -K_D\frac{\partial c}{\partial x} \qquad (C.3)$$

The Equation C.2 can now be written as:

$$\frac{\partial c}{\partial t} + U\frac{\partial c}{\partial x} - K_D\frac{\partial^2 c}{\partial x^2} = 0 \qquad (C.4)$$

Where
A	= cross sectional area (m^2)
Q	= water discharge (m^3/s)
K_D	= diffusion coefficient (m^2/s)

U = advection velocity (m/s)

c = concentration of sediment (m³ of sediment/m³ of water)

In unsteady flows the convective/advective term and diffusion term are solved separately as described below:

Solution of the advection term

The advection term is solved by the Method of Characteristics using Holly-Preissmann Scheme.

Partial differential equation of advection is written as:

$$\frac{\partial c}{\partial t} + U \frac{\partial c}{\partial x} = 0 \qquad\qquad (C.5)$$

This partial differential equation can be written as an ordinary differential equation (ODE) along the characteristic curve $dx/dt = u(x,t)$ as:

$$\frac{dc}{dt} = 0 \qquad\qquad (C.6)$$

Method of Characteristics

The computational grid in the Figure C.2 illustrates the solution of the ordinary differential equation (ODE) $dc/dt = 0$ along the characteristic curve $dx/dt = u$. Here the value of $c_i^{j+1} = c(x_i, t_{j+1})$ along the characteristic curve is determined that starts at position $x = \xi$ at time level j. Integrating $dc/dt = 0$ along the trajectory illustrated in the Figure C.2 produces $c_i^{j+1} = c_\xi$.

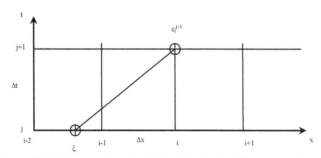

Figure C.2. Computational grid for Method of Characteristics

To determine C_ξ the Holly-Preissmann scheme is used in which a third-order interpolating polynomial is constructed which is based on the information from two points $(i-1, j)$ and (i, j).

This polynomial is written as:

$$c(r) = A_1 + A_2 r + A_3 r^2 + A_4 r^3 \tag{C.7}$$

where

$$r = \frac{x_i - \xi}{\Delta x_i} \tag{C.8}$$

and the A's are constant values. The parameter r is illustrated in the Figure C.3 showing the characteristic line in a cell of the computational grid. (for a constant value of u, the characteristic curve is a straight line and the parameter r is also the Courant number of the computational grid: $r = u.\Delta t / \Delta x$).

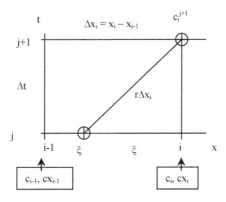

Figure C.3. Illustration of parameter r in a computational grid

In order to evaluate the constants of the cubic polynomial in Equation C.7, it is needed to provide four values of (r, c). However, since there are only two reference points where c is known $(i\text{-}1, j)$ and (i, j), it is needed to include also the derivative values $cx = \partial c / \partial x$ at those reference points. The term cx can be written in terms of r as follows:

$$cx = \frac{\partial c}{\partial x} = \frac{dc}{dr}\frac{dr}{dx} = c'(r)(-1/\Delta x_i) \tag{C.9}$$

Since, in the context of the calculation cell, $x = \xi$, the result $dr/dx = dr/d\xi = -1/\Delta x_i$ follows from Equation C.8.

Then the equation for derivative cx becomes:

$$cx(r) = -c'(r)/\Delta x_i = -(A_2 + 2A_3r + 3A_4r^2)/\Delta x_i \qquad (C.10)$$

Figure C.3 illustrates the fact that the values c_{i-1}, cx_i, c_i, cx_i are known. These values correspond to $r = 1$ for $x = x_{i-1}$ and $r = 0$ for $x = x_i$. Thus, replacing the values of c, cx and r, for points x_{i-1} and x_i, into Equations C.7 and C.9, produce a system of linear equations whose solution is given by:

$$
\begin{aligned}
A_1 &= c_i \\
A_2 &= -\Delta x_i cx_i \\
A_3 &= (2cx_i + cx_{i-1})\Delta x_i + 3(c_{i-1} - c_i) \\
A_4 &= -(cx_i + cx_{i-1})\Delta x_i + 2(c_i - c_{i-1})
\end{aligned}
\qquad (C.11)
$$

Replacing these constants into Equation C.7 and collecting terms allow to re-write the equation as:

$$c(r) = a_1 c_{i-1} + a_2 c_i + a_3 cx_{i-1} + a_4 cx_i \qquad (C.12)$$

Where

$$
\begin{aligned}
a_1 &= r^2(3 - 2r) \\
a_2 &= 1 - r^2(3 - 2r) = 1 - a_1 \\
a_3 &= r(3r - 2)\Delta x_i \\
a_4 &= (r - 1)(3r - 1)
\end{aligned}
\qquad (C.13)
$$

Also, Equation C.10 becomes:

$$cx(r) = b_1 c_{i-1} + b_2 c_i + b_3 cx_{i-1} + b_4 cx_i \qquad (C.14)$$

Where

$$
\begin{aligned}
b_1 &= 6r(1 - r)/\Delta x_i \\
b_2 &= 6r(r - 1)/\Delta x_i = -b_1 \\
b_3 &= r(3r - 2)\Delta x_i \\
b_4 &= (r - 1)(3r - 1)
\end{aligned}
\qquad (C.15)
$$

Initial conditions

The initial conditions for the concentration and its derivative in x are simply values of $c_i^j = 1 = c(x_i, t_1)$ and $cx_i^{j+1} = cx(x_i, t_1)$. These values should be known.

Boundary conditions

Only the upstream boundary condition is needed. The simplest case is when values of both $c_1(t) = c_1^j = c(x_1, t)$ and $cx_1(t)=cx_1^j = \partial c / \partial x \mid_{1,t}$ are provided. Alternatively, only the values of $c_1(t)$ are provided.

Solution of diffusion term

One dimensional diffusion equation in x-direction can be written as:

$$\frac{\partial c}{\partial t} = K_D \frac{\partial^2 c}{\partial x^2} \tag{C.16}$$

In fractioned approach the contributions to the variation in c by advection and diffusion are solved sequentially. As a first step, the advection equation (Eq. C.2) is solved, so as to give the c-values at $(n+1)$ Δt resulting from transport alone, while as a second step the equation:

$$\frac{\partial c}{\partial t}\Big|_{diffusion} = \frac{\partial c}{\partial t} - \frac{\partial c}{\partial t}\Big|_{advection} = K_D \frac{\partial^2 c}{\partial x^2} \tag{C.17}$$

is solved, where K_D is diffusion coefficient, giving complete solution of the advection and diffusion together.

Smith (1978) describes the Crank-Nicolson finite difference method for solving the one dimensional diffusion equation, which is given below:

$$\frac{c_i^{j+1} - c_i^j}{\Delta t} = \frac{K_D}{2\Delta x^2}\left((c_{i+1}^{j+1} - 2c_i^{j+1} + c_{k-1}^{j+1}) + (c_{i+1}^j - 2c_i^j + c_{i-1}^j)\right) \tag{C.18}$$

giving

$$-rc_{i-1}^{j+1} + (2+2r)c_i^{j+1} - rc_{i+1}^{j+1} = rc_{i-1}^j + (2-2r)c_i^j + rc_{i+1}^j \tag{C.19}$$

Where

$$r = \frac{K_D \Delta t}{\Delta x^2} \tag{C.20}$$

This is a *tri-diagonal* problem. The concentration at c_i^{j+1} can be solved by using the *tri-diagonal* matrix algorithm.

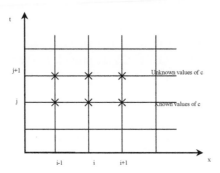

Figure C.4: Crank-Nicolson computational grid for 1-D diffusion equation

In general, the left side of the Equation C.18 contains three unknown and the right side three known, pivotal values of c. If there are N internal mesh points along each time row then for $j = 0$ and $i = 1, 2,3, ...N$, Equation C.18 gives N simultaneous equations for the N unknown pivotal values along the first time row in terms of known initial and boundary values. Similarly $j = 1$ expresses N unknown values of u along the second time row in terms of the calculated values along the first etc.

D. Modified Einstein Procedure for computation of total sediment load from field measurements

Modified Einstein Procedure (MEP) is used to compute total sediment load from the field measurements. The water and sediment measurements include discharge measurements and suspended sediment load measurements. MEP is described here as given by Simons and Senturk (1992). In discharge measurements the depth and means flow velocity are measured at each vertical in a cross-section, from which the total discharge Q and cross-sectional area are calculated. In suspended sediment measurements the sediment concentration C_s is measured at each, or at least several, of the verticals in which mean velocity is measured. The concentration C_s includes all grain sizes, wash load as well as bed-material load and is defined by:

$$c_s \int_{a'}^{d} u dy = cudy \tag{D.1}$$

Where
a'	= distance from the bed to the sampler inlet tube (m)
u	= flow velocity (m/s)
c_s	= measured suspended sediment concentration (ppm)
d	= flow depth (m)
c	= sediment concentration by volume (m³ of sediment / m³ of water)
dy	= flow depth increments (m)

The region below a is the unsampled zone of the cross-section and the part of the cross-section above a is called the sampled zone.

The modified Einstein Procedure (MEP) developed by Colby and Hembree (1955), the Colby (1957) method and the Toffaletti (1969) method are generally used to estimate the total sediment discharge, once suspended load samples and other required data have been collected. In this study the MEP method has been used to compute the total sediment load.

To compute the total sediment load by MEP the data needed are water discharge, mean flow velocity, cross-sectional area, channel width, mean value of the depths at verticals where suspended sediment samples were taken, the measured sediment discharge concentration, the size distribution of measured load, the size distribution of the bed material at the cross-section and the water temperature.

The suspended-load discharge is computed then in the various size fractions per unit width in the sample zone of the cross-section and is given by:

$$q_s' = \sum_i q_{si}' = \frac{C_s \gamma Q'}{b} \tag{D.2}$$

The relationship between Q' and Q is given by:

$$\frac{Q'}{Q} = \frac{\int_{a'}^{d} u \, dy}{\int_{0}^{d_v} u \, dy}$$

(D.3)

Where

q_s,   = suspended sediment discharge per unit width
Q' = water discharge in the sampled zone (m³/s)
Q = total water discharge (m³/s)
q_{si},   = size fractions of suspended sediments per unit width
γ = specific weight of water (T/m³)
b = water surface width (m)
d_v = mean value of depths at verticals of sediment sampling

If the point velocity is taken as average velocity of the following form:

$$u = U + \frac{u_*}{\kappa} + \frac{2.30 u_*}{\kappa} \log \frac{y}{d_u}$$

(D.4)

Where

u = velocity at a point (m/s)
U = mean flow velocity (m/s)
κ = universal constant of Vón Kármán
y = flow depth (m)
d_u = depth of the point of velocity measurement (m)

Then

$$\frac{Q'}{Q} = (1 - E') - 2.3 \frac{E' \log E'}{P_m - 1}$$

(D.5)

where $E' = a'/d_v$, and for practical purposes:

$$P_m \cong 2.3 \log \frac{30.2 x d}{D_{65}}$$

(D.6)

Where x is a correction factor that compensates for conditions where the channel bed is not fully rough and is a function of k_s/δ. The quantity of x is found from Figure D.1 with the help of following equation:

$$\frac{U}{u_*} = 5.75 \log \left(12.27 \frac{R'}{k_s} x \right)$$

(D.7)

Where
q_{si}' = size fraction of suspended load per unit width in sample zone
q = actual water discharge per unit width (m²/s)
R' = related to skin friction (m)
i_s = size distribution of suspended sediment sample
u_* = shear velocity due to grain roughness (m/s)
d = flow depth (m)
D_{65} = particle size for which 65% of the sediment mixture is finer (m)
k_s = equivalent sand grain roughness (m)
δ = thickness of laminar sublayer (m)

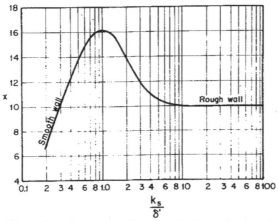

Figure D.1. Correction factor in the logarithmic velocity distribution

$$q_{si}' = i_s \gamma C_s q \left[(1 - E') - 2.3 \frac{E' \log E'}{P_m - 1} \right] \qquad (D.8)$$

The bed load for a given size fraction per unit width of channel $i_{Bw}q_{bw}$ is computed next. The intensity of shear on the particles is ψ is calculated from the following equations:

$$\psi = \frac{\left(\frac{\gamma_s}{\gamma} - 1 \right) D_{35}}{(S_f R')} \qquad (D.9)$$

and for geometric means size of the particles:

$$\psi = \frac{0.4 \left(\frac{\gamma_s}{\gamma} - 1 \right) D_i}{(S_f R')} \qquad (D.10)$$

Where

ψ	= shear intensity parameter
γ_s	= specific weight of sediment (T/m³)
S_f	= energy slope (m/m)
D_i	= geometric mean particle size (m)
D_{35}	= particle size for which 35% of the sediment mixture is finer (m)
γ	= specific weight of water (T/m³)

The larger ψ value is used to find ϕ_* from Figure D.2 where ψ replaces ψ_*.

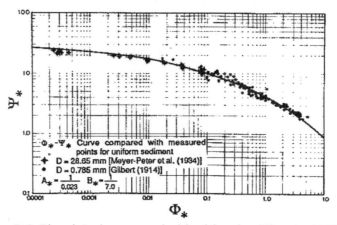

Figure D.2. Einstein's ϕ_* versus ψ_* bed-load function (Einstein, 1950)

Then from the definition of ϕ_*:

$$\frac{p}{1-p} = (A_*)\frac{i_{Bw}}{i_{bw}}(\phi) = A_*\phi_* \tag{D.11}$$

Where $A_* = A1A3/A2\lambda$

$$\phi_* = \frac{i_{Bw}q_{bw}}{i_{bw}g\rho_s}\sqrt{\frac{\rho}{\rho_s-\rho}}\sqrt{\frac{1}{gD^3}} \tag{D.12}$$

and $q_{bw} = q_b$

It gives the following equation:

$$i_{Bw}q_{bw} = \frac{1}{2}\phi_* i_{bw}\gamma_s\sqrt{\frac{\gamma_s}{\gamma}-1}\sqrt{gD_i^3} \tag{D.13}$$

Where

A_1, A_2, A_3 = constants
p = probability for a particle to be eroded at any one spot
i_{Bw} = fraction of the total sediment load for a given sediment size
i_{bw} = fraction of bed sediment of a given sediment size
ϕ = transport rate function
q_{bw} = bed load discharge by weight per unit width per unit time
g = acceleration due to gravity (m/s^2)
ρ = density of fluid (kg/m^3)
ρ_s = density of sediment (kg/m^3)
λ = porosity of the channel bed

The term ϕ_* is arbitrarily divided by a factor of two to fit the observed channel data more closely.

The next step is to compute the suspended load exponent z_i' by trial and error for each size fraction. If the quantity q_{si}' can expressed by:

$$q_{si}' = \int_{a'}^{d_y} C_i u \, dy \qquad (D.14)$$

then using the following procedure:

$$\frac{q_{si}'}{i_{Bw} q_{bw}} = \frac{I_1(E, z')}{J_1(E, z')} \left[P_m J_1(E', z') + J_2(E', z') \right] \qquad (D.15)$$

where

$$J_1(E', z') = \int_{E'}^{1} \left(\frac{1-y}{y} \right)^{z'} dy \qquad (D.16)$$

$$J_2(E', z') = \int_{E'}^{1} \left(\frac{1-y}{y} \right)^{z'} \ln y \, dy \qquad (D.17)$$

and

$$I_1(E, z') = 0.216 \frac{E^{z-1}}{(1-E)^z} \int_{E}^{1} \frac{(1-y)^z}{y} dy \qquad (D.18)$$

and $E = a/d$, $z = w_f/0.4u_*'$, $z' = w_{fi}/0.4u_*'$
Where

a   = reference level, limit height between suspended and bed load (m)
d   = water depth (m)

Let $I = I(E, z')$; $J = J(E, z')$; $I' = I(E', z')$, $J' = J(E', z')$

The above equation now becomes:

$$\frac{Q'_{si}}{i_{Bw}Q_{bw}} = \frac{Bq'_{si}}{B(i_{Bw}q_{bw})} = \frac{I_1}{J_1}(P_m J'_1 + j'_2)$$
(D.19)

The values of I_1, I_2, J_1 and J_2 can be obtained from the Figures D.3, D.4, D.5 and D.6. In the above equation the quantity $(Q_{si}'/i_{Bw}Q_{bw})$ for each size fraction is known. The value of z_i' can then be found by trial and error so that the above equation is satisfied. In this way, the z_i' value for each size fraction can be solved. However, Colby and Hembree (1955) found that the value of z_i' can be related to the fall velocities of the sediment by:

$$\frac{z_i'}{z_1'} = \left(\frac{w_{f\,i}}{w_{f_1}}\right)^{0.07}$$
(D.20)

where z_1' is obtained by solving Equation D-15 for the dominant grain size, and w_{fi} is the fall velocity of a sediment grain size D_i.

The total sediment discharge through the cross-section Q_{Ti} for a size fraction is then computed from:

$$Q_{Ti} = Q'_{si}\frac{P_m J_1 + J_2}{P_m J'_1 + J'_2}$$
(D.21)

for the ranges of fine particle sizes and the relation:

$$Q_{Ti} = i_{Bw}Q_{bw}(P_m I_1 + I_2 + 1)$$
(D.22)

is utilized to evaluate the transport of coarse particle sizes. The units of Q_{Ti} are the dry weight per unit of time. Any consistent system of units may be used.

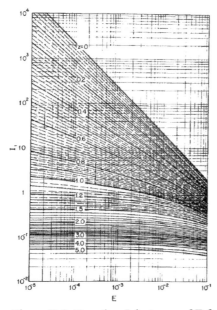

Figure D.3. Function I_1 in terms of E for values of z (Einstein, 1950)

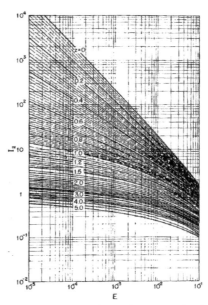

Figure D.4. Function I_2 in terms of E for values of z (Einstein, 1950)

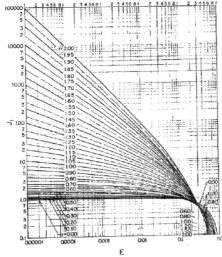

Figure D.5. Integral J_1 in terms of E and z' (Colby and Hembree, 1955)

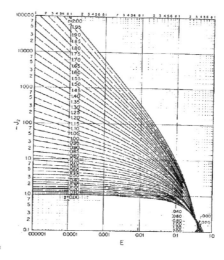

Figure D.6. Integral J_2 in terms of E and z' (Colby and Hembree, 1955)

E.   Crop water requirements and Crop Based Irrigation Operations

E.1. Crop Based Irrigation Operation (CBIO)

Higher irrigation deliveries give numerous benefits to the farmers by providing ample amount of water for cultivating high delta and high value crops. But it results in waterlogging due to over irrigation in the long run. To deal with this situation a canal operation plan, the Crop Based Irrigation Operations (CBIO), was prepared. This operation plan aims at controlling the groundwater recharge and saving irrigation water by limiting extra irrigation deliveries to the command area.

E.1.1. Concept

CBIO is a canal operation strategy for controlling irrigation water supplies in order to keep a balance between the water demands and supplies and thus preventing crop failure, saving water and above all controlling groundwater recharge.

CBIO basically limits irrigation supplies to the command area when crop water requirements are low, contrary to the traditional supply based irrigation operations under which full discharges are delivered to the command area. CBIO is a modification of the supply based irrigation operations and it is very effective in canal operation having high water allowance like 0.5 to 0.7 litres/second/hectare (l/s/ha).

CBIO is applied when crop water requirements are below the design water allowance. A curvilinear relationship is given in Figure E.1, in which the typical variations in crop water requirements has been shown.

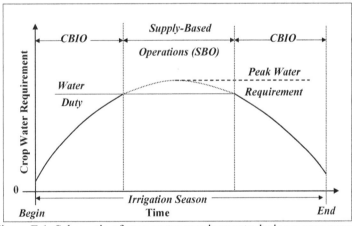

Figure E.1. Schematic of crop water requirements during a crop season

The approach is to run the system on full supply and stop the supplies for few days, appropriate to meet the crop water demand. The appropriate way is to close and open the offtake for a seven days cycle e.g. 7 days close and 7 days open. This rotation is in compliance with the prevailing local weekly *warabandi* system and hence none of the shareholder would lose its *warabandi* turn.

E.1.2. Preparation of hydrograph for canal water deliveries

First of all the crop water requirements are determined and then the hydrograph is prepared according to the demand.

E.1.3. CBIO duration

Warabandi is a predominant water distribution system at tertiary level in Pakistan, in which a farmer has the full watercourse supply for a duration of time in accordance to landholding in the culturable command area. Most frequently warabandi has a duration of one week in IBIS. So the CBIO duration is multiples of weeks so that a farmer does not miss his *warabandi* turn more than a week.

E.1.4. Canal cycle and duration Options

The duration of a cycle period is the sum of the time for which the secondary canals are open plus closed. The closure periods are not more than one *warabandi* turn. This implies that the closed period during each cycle duration are not more that one week. The minimum cycle period for a channel is two weeks, with one week open and one week close. This is followed by a cycle of three weeks with two week open and one week close. Further the cycle duration can be four weeks, five weeks and a maximum of six weeks with five week open and one week close.

E.1.5. Canal cycle discharge rates

A particular fraction of the discharge is supplied under each cycle duration e.g. under the cycle of two weeks with one week open and one week close a 50 percent of the design discharge, or full supply discharge (FSD) is supplied. This relationship can be found as follows:

$$Canal\ Cycle\ Discharge = \frac{Open\ Period}{Cycle\ Duration} * FSD \qquad\qquad (E.1)$$

or

$$Canal\ Cycle\ Discharge = Discharge\ Fraction * FSD \qquad\qquad (E.2)$$

The results are shown in Table E.1.

Table E.1: Discharge fraction for various cycle duration

Open Period (week)	Closed Period (week)	Cycle Duration (week)	Discharge Fraction
1	1	2	0.50
2	1	3	0.67
3	1	4	0.75
4	1	5	0.80
5	1	6	0.83

E.1.6. Adjusting canal cycle discharge rates

There is another degree of flexibility in implementing CBIO. The discharge rates can be either increased or decreased by a nominal amount. Most secondary canals can be operated at a discharge rate 10% greater than FSD. Generally, it is most desirable to increase the discharge rate for the shortest cycle durations; primarily to reduce the amount of sediment deposition in the canal. Consequently, first preference would be given to increase the discharge rate for the shortest cycle duration. At the same time, discharge rates can be reduced, but this is avoided. However, this may occur where there is a water supply constraint. If this is the case, then preference would be given to reduce the water duty for supply-based operations (SBO), followed by the longer cycle durations being used during crop based irrigation operations (CBIO).

E.1.7. Distributary Rotation Schedule

Rotation among reaches. If there are sufficient cross-regulators along the canal, then it may be feasible to rotate among the reaches defined by these cross-regulators. Then, all of the distributaries between two cross-regulators will either be open or closed. For a cycle duration of two, with one time period open and one time period closed, the rotation would be done among two reaches. Likewise, for a cycle duration of three, with two time open and one time period close, three clusters of reaches would be used for developing the rotation schedule for each reach. The number of reaches along a canal is not likely to be some multiple of two, three or four. Thus, there may be a need to: (1) group two reaches together, usually selecting the reaches having the lowest discharge requirements for combining with one another or combining with a reach having an intermediate discharge requirement; (2) operate one or two reaches independently without combining with any other reaches; or (3) operate one or two reaches where distributaries within the reach are rotated as described below.

Rotation Among Distributaries. CBIO is implemented by rotating water deliveries among the distributaries in a reach, or developing clusters of distributaries scattered along the full length of the canal is used to develop clusters of distributaries. For a cycle duration of two, with one time period open and one time period closed, the distributaries in a reach will have to be clustered into two groups of nearly total discharge requirement. Similarly, a cycle duration of three, with one time period closed, requires that the distributaries in a reach be clustered into three groups of fairly equal total discharge requirement. A cycle duration of four, with one time period closed, requires four clusters of distributaries in each reach. The alternative is to define these clusters along the full length of the canal, which has the advantage of minimizing the variation of total discharge for each cluster.

Schedule adjustment for lag times. One time period in a cycle duration represents the time required to rotate the water supply in a tertiary irrigation channel to all of the farmers. However, when the gates of a secondary canal are opened, a tertiary channel located nearby will begin to receive water in just a few minutes, while the last tertiary channel at the tail of the secondary canal will not receive any water until hours later. Also, when the secondary canal gates are closed, the discharge rate entering the head tertiary canal will rapidly decline, while the tail tertiary canal will continue to receive water. A perfect solution would be that the lag times to the tail tertiary canal are identical when opening the secondary canal gates and when closing them.

E.2. Crop Water Requirements

E.2.1. Concept:

Crop water requirement or crop evapotranspiration is the total amount/depth of water required by the crop for its maturity.

E.2.2.   Calculation of crop evapotranspiration (ET$_c$)

Crop evapotranspiration is the product of the potential evapotranspiration and the crop coefficient.

$$ET_c = ET_o * k_c \qquad (E.3)$$

Where
ET_c = Crop evapotranspiration (mm)
ET_o = Potential evapotranspiration (mm)
K_c = Crop coefficient

E.2.3. Potential (Reference) evapotranspiration (ET$_o$)

The potential evapotranspiration or reference evapotranspiration is the rate of the evapotranspiration from a large area, covered by a green grass (reference crop) such as alfalfa, 8 to 15 cm tall which grows actively, completely shades the ground and which is not short of water.

ETo is dependant upon the climatic conditions. It depends essentially on the energy available to evaporate moisture and carry it away and can be calculated from the climatological data like: solar radiation, minimum and maximum temperatures, relative humidity and wind speed etc.

E.2.4. Calculation of ETo

A number of methods are available for this purpose like:

   i.   Blanney Criddle
   ii.   Radiation
  iii.   Penman
  iv.   Pan Evaporation
   v.   Modified Penman-Montieth

A computer model CropWAT was used in this study to calculate the crop water requirements. This model has been developed by Food and Agriculture Organization (FAO) of the United Nations. The CropWAT model uses the Modified Penman-Montieth equation for this purpose, which is given by (Allen et al., 1998):

$$ET_o = \frac{0.408\Delta(R_n - G) + \gamma\dfrac{900}{T+273}u_2(e_s - e_a)}{\Delta + \gamma(1 + 0.34u_2)} \qquad (E.4)$$

Where

ET_o	= reference evapotranspiration (mm/day)
R_n	= net radiation at the crop surface (MJ/m²/day)
G	= Soil heat flux density (MJ/m²/day)
T	= mean daily air temperature at 2 m height (°C)
u_2	= wind speed at 2 m height (m/s)
e_s	= saturation vapour pressure (kPa)
e_a	= actual vapour pressure (kPa)
$e_s\text{-}e_a$	= saturation vapour pressure deficit (kPa)
Δ	= slope vapour pressure curve (kPa/oC)
γ	= psychrometric constant (kPa/oC)

E.2.5   Data required for the calculation of ET₀

This data can be categorized in two forms

General Data

Altitude, Longitude and Latitude of the station, for which the ET_o is calculated.

Climate Data

Climate data is the base for calculation of the reference evapotranspiration. CropWat Model needs very common parameters and using these it derives itself the other complicated parameters and finally calculates the ET_o. The data required are as follows:

    i.   Mean minimum temperature
    ii.   Mean maximum temperature
    iii.   Relative humidity
    iv.   Sun shine hours and
    v.   Wind speed

E.2.6. Sources of climate data

Climate data representative of Maira-PHLC command area was collected from the following meteorological stations, nearby Swabi Irrigation Division.

    i.   Sugar Crops Research Institute (SCRI), Mardan
    ii.   Cereal Crops Research Institute (CCRI), Pir Sabaq
    iii.   Water and Power Development Authority (WAPDA), Tarbela

Table E.2. Climate data of the study area (30 years average)

Month	Max. Temp. °C	Min. Temp. °C	Air humidity (%)	Wind speed (km/day)	Daily sunshine hours (hrs)
Jan	18.6	3.5	68	16.4	8
Feb	20.2	5.7	65	25.9	8.7
Mar	25.1	10.0	66	25.9	9.5
Apr	32.4	15.3	56	26.8	10.6
May	38.3	20.5	45	25.9	12.2
Jun	42.2	24.4	40	36.3	12.8
Jul	39.3	25.2	62	54.4	10.4
Aug	36.5	24.2	70	42.3	10.3
Sep	36.1	21.4	61	24.2	10.8
Oct	32.7	15.6	60	13.0	10.3
Nov	26.5	8.7	67	8.6	9.0
Dec	20.3	4.3	72	8.6	7.7

Then the ET_o was determined for the Lower Machai Branch Canal and the Maira Branch Canal as given in Table E.3.

Table E.3. ETo of Maira-PHLC command area

Month	ETo (mm/day)
January	1.02
February	1.74
March	2.84
April	4.25
May	5.44
June	6.23
July	5.93
August	5.22
September	4.22
October	2.64
November	1.35
December	0.82

E.2.7. Crop data required

After calculating the potential evapotranspiration, the next step is to use crop data for calculating crop evapotranspiration (or crop water requirements). The crop data required are as follows;

   i.   Crop type
   ii.   Crop duration (days)
   iii. Growth stages with duration (days)
   iv.   Sowing dates
   v.   Harvesting dates

vi.   Crop K_c values

Most of the data is already available in the model by default only cropping pattern, sowing and harvesting dates need to be entered. Then with the help of this data CropWat Model calculates the depth (in mm) of water required to mature the crops given. The cropping pattern of the study area is given in the Figure E.2.

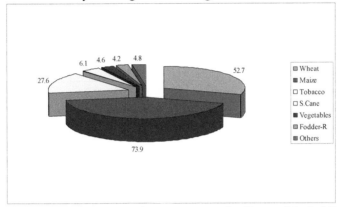

Figure E.2. Cropping pattern (cropped area percentage) of Lower Machai Branch Canal and Maira Branch Canal

E.2.8   Crop water requirements for the command area downstream of RD 242

The daily crop water requirement for the command area below RD 242 is shown in Figure E.3.

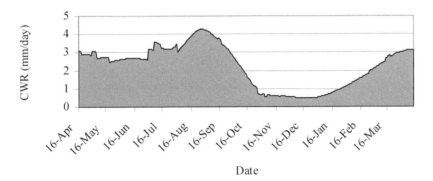

Figure E.3. Daily crop water requirements d/s RD 242

E.3. Preparation of the Crop Based Irrigation Operations Schedules

Crop Based Irrigation Operations (CBIO) schedule is prepared on the basis of crop water requirements (CWR) of the canal command area. A computer model CBIO is used for preparing schedules. The arrangement of the CBIO Schedule preparation by using CBIO Model is given as under.

Table E.4.A. Weekly CBIO Schedule for Kharif 2007-08

Days	Sarbandi Minor	Old indus	Dagi Disty	Yar Hussain Minor	Sadri Minor	Ghazikot Minor	Neknam Minor	Sudher Minor	Yaqubi Minor	Sardchina Minor	Dault Minor	Gumbat-I Minor	Gumbat-II Minor	Qasim-I Minor	Qasim-II Minor	Toru Minor	Pirsabaq Disty	Chowki Disty
16-22/April	O	C	O	O	O	O	O	O	O	O	C	O	O	O	O	O	O	C
23-29/April	O	O	O	O	O	C	O	O	C	O	O	C	C	O	O	C	C	O
30/April-06/May	C	O	C	C	C	O	C	C	O	C	O	O	O	C	C	O	O	O
07-13/May	O	C	O	O	C	O	O	O	O	O	O	O	O	O	O	O	O	C
14-20/May	O	O	O	O	O	O	O	O	C	O	O	C	O	O	O	O	C	O
21-27/May	O	O	C	O	O	C	O	O	O	O	O	O	C	C	O	C	O	O
28/May-03/Jun	C	O	O	C	O	O	C	C	O	C	C	O	O	O	C	O	O	O
04-10/June	O	O	O	O	O	O	O	O	O	O	O	C	O	O	O	O	O	C
11-17/Jun	O	O	O	O	O	O	O	O	C	O	O	O	O	O	O	O	C	O
18-24/June	O	O	C	O	O	O	O	O	O	O	O	O	C	O	O	C	O	O
25/June-01/July	O	O	O	C	O	C	O	O	O	O	C	O	O	C	O	O	O	O
02-08/July	C	C	O	O	C	O	C	C	O	C	O	O	O	O	O	C	O	O
09-15/July	O	O	O	O	O	O	O	O	O	O	O	O	O	O	O	O	O	C
16-22/July	O	O	O	O	O	O	O	O	C	O	O	O	O	O	O	O	C	O
23-29/July	O	O	C	O	O	O	O	O	O	O	O	O	O	O	O	C	O	O
30/July-05/Aug	O	O	O	O	O	C	O	O	O	O	C	O	C	O	O	O	O	O
06-12/Aug	O	C	O	C	O	O	O	O	O	O	O	O	O	C	C	O	O	O
13-19/Aug	C	O	O	O	C	O	C	C	O	C	O	C	O	O	O	O	O	O
20/Aug-07/Sep	O	O	O	O	O	O	O	O	O	O	O	O	O	O	O	O	O	O
10-16/Sep/05	O	O	O	O	O	O	O	O	O	O	O	O	O	O	O	O	O	O
17-23/Sep	O	O	O	O	O	O	O	O	C	O	O	O	O	O	O	O	C	O
24-30/Sep	O	O	C	O	O	O	O	O	O	O	O	O	O	O	O	C	O	O
01-07/Oct	O	O	O	O	O	C	O	O	O	O	C	O	C	O	O	O	O	O
08-14/Oct	O	C	O	C	O	O	O	O	O	O	O	O	O	C	C	O	O	O
15-22/Oct	C	O	O	O	C	O	C	C	O	C	O	C	O	O	O	O	O	O

Table E.4.B. Weekly CBIO Schedule for Kharif 2007-08

	Sarbandi Minor	Old indus	Dagi Disty	Yar Hussain Minor	Sadri Minor	Ghazikot Minor	Neknam Minor	Sudher Minor	Yaqubi Minor	Sardchina Minor	Dault Minor	Gumbat-I Minor	Gumbat-II Minor	Qasim-I Minor	Qasim-II Minor	Toru Minor	Pirsabaq Disty	Chowki Disty
15-21/Oct	C	O	O	O	C	O	C	C	O	C	O	C	O	O	O	O	O	O
22-28/Oct	O	O	O	O	O	O	O	O	O	O	O	C	O	O	O	O	O	C
29/Oct-04/Nov	O	O	O	O	O	O	O	O	C	O	O	O	O	O	O	O	C	O
05-11/Nov	O	O	C	O	O	O	O	O	O	O	O	O	C	O	O	C	O	O
12-18/Nov	O	O	O	C	O	C	O	O	O	O	C	O	O	C	O	O	O	O
19-25/Nov	C	C	O	O	C	O	C	C	O	C	O	O	O	O	C	O	O	O
26/Nov-02/Dec	O	O	C	O	O	O	O	O	C	O	C	C	O	O	O	O	O	C
03-09/Dec	C	O	O	C	C	C	C	C	O	C	O	C	C	C	C	C	C	O
10-16/Dec	O	C	C	O	O	O	O	O	C	O	C	C	O	O	O	O	O	C
17-23/Dec	C	O	O	C	C	C	C	C	O	C	O	O	C	C	C	C	C	O
24-31/Dec	O	C	C	O	O	O	O	O	C	O	C	C	O	O	O	O	O	C
1/Jan-3/Feb	C	C	C	C	C	C	C	C	C	C	C	C	C	C	C	C	C	C
04-10/Feb	O	C	C	O	O	O	O	O	C	O	C	C	O	O	O	O	O	C
11-17/Feb	C	O	O	C	C	C	C	C	O	C	O	O	C	C	C	C	C	O
18-24/Feb	O	C	C	O	O	O	O	O	C	O	C	C	O	O	O	O	O	C
25/Feb-03/Mar	C	O	O	C	C	C	C	C	O	C	O	O	C	C	C	C	C	O
04-10/Mar	O	C	C	O	O	O	O	O	C	O	C	C	O	O	O	O	O	C
11-17/Mar	C	O	O	C	C	C	C	C	O	C	O	O	C	C	C	C	C	O
18-24/Mar	O	C	O	O	C	O	O	O	O	O	O	O	O	O	O	O	O	C
25-31/Mar	O	O	O	O	O	O	O	O	C	O	O	C	O	O	O	O	C	O
01-07/Apr	O	O	C	O	O	C	O	O	O	O	O	O	C	C	O	C	O	O
08-14/Apr	C	O	O	C	O	O	C	C	O	C	C	O	O	O	C	O	O	O

F.   Sediment inflow at Machai Branch headworks

Date	Day	Flow m³/s	Concentration kg/m³	Particle size % > 0.062	% < 0.062
03/04/2007	Tuesday	40.4	0.127	7	93
01/05/2007	Tuesday	45.5	0.287	9	91
21/05/2007	Monday	44.8	0.201	8	92
23/05/2007	Wednesday	44.8	0.199	14	86
25/05/2007	Friday	44.8	0.163	19	81
27/05/2007	Sunday	44.8	0.171	3	97
29/05/2007	Tuesday	44.8	0.174	2	98
31/05/2007	Thursday	46.2	0.089	5	95
02/06/2007	Saturday	46.2	0.105	2	98
04/06/2007	Monday	46.2	0.215	12	88
06/06/2007	Wednesday	46.2	0.123	1	99
08/06/2007	Friday	46.2	0.101	8	92
10/06/2007	Sunday	46.2	0.251	19	81
12/06/2007	Tuesday	50.5	0.160	7	93
14/06/2007	Thursday	50.5	0.085	7	93
16/06/2007	Saturday	50.5	0.421	10	90
20/06/2007	Wednesday	48.1	0.130	17	83
22/06/2007	Friday	49.3	0.209	16	84
24/06/2007	Sunday	49.3	0.231	10	90
26/06/2007	Tuesday	49.3	0.745	23	77
28/06/2007	Thursday	52.7	0.737	11	89
28/06/2007	Thursday	52.7	0.842	12	88
30/06/2007	Saturday	28.4	0.337	7	93
02/07/2007	Monday	28.4	0.439	25	75
04/07/2007	Wednesday	31.7	0.183	23	77
06/07/2007	Friday	31.7	0.226	10	90
08/07/2007	Sunday	31.7	0.498	15	85
11/07/2007	Wednesday	47.7	0.182	12	88
12/07/2007	Thursday	47.4	0.183	14	86
13/07/2007	Friday	46.2	1.912	17	83
14/07/2007	Saturday	47.4	0.151	11	89
15/07/2007	Sunday	47.2	0.094	14	86
16/07/2007	Monday	47.4	0.154	19	81

Date	Day	Flow m^3/s	Concentration kg/m^3	Particle size % > 0.062	% < 0.062
17/07/2007	Tuesday	46.2	0.152	17	83
18/07/2007	Wednesday	47.2	0.142	12	88
19/07/2007	Thursday	47.4	0.111	14	86
20/07/2007	Friday	47.4	0.079	15	85
21/07/2007	Saturday	47.5	3.514	21	79
22/07/2007	Sunday	41.3	0.654	16	84
23/07/2007	Monday	41.3	0.105	16	84
24/07/2007	Tuesday	42.1	1.019	10	90
25/07/2007	Wednesday	41.3	0.106	14	86
26/07/2007	Thursday	46.7	0.138	17	83
27/07/2007	Friday	42.1	0.141	11	89
28/07/2007	Saturday	42.1	0.080	12	88
29/07/2007	Sunday	46.7	0.087	10	90
30/07/2007	Monday	46.7	0.153	14	86
31/07/2007	Tuesday	47.4	0.199	13	87
01/08/2007	Wednesday	47.4	1.671	19	81
02/08/2007	Thursday	47.4	0.261	19	81
03/08/2007	Friday	47.9	0.245	17	83
04/08/2007	Saturday	47.4	0.236	21	79
05/08/2007	Sunday	47.4	0.231	20	80
06/08/2007	Monday	47.4	0.263	21	79
07/08/2007	Tuesday	46.2	0.084	22	78
08/08/2007	Wednesday	47.2	0.278	26	74
09/08/2007	Thursday	44.8	0.228	21	79
10/08/2007	Friday	42.1	0.216	20	80
11/08/2007	Saturday	40.5	0.401	27	73
12/08/2007	Sunday	48.1	1.557	29	71
20/08/2007	Monday	48.1	0.610	27	73
22/08/2007	Wednesday	40.5	0.033	14	86
24/08/2007	Friday	38.4	0.130	21	79
26/08/2007	Sunday	39.8	0.147	18	82
28/08/2007	Tuesday	37.0	0.103	17	83
30/08/2007	Thursday	28.4	0.173	18	82
01/09/2007	Saturday	27.4	0.217	19	81
03/09/2007	Monday	31.7	0.178	17	83
05/09/2007	Wednesday	37.0	0.382	18	82
07/09/2007	Friday	50.2	0.347	16	84
09/09/2007	Sunday	46.7	0.264	12	88

Date	Day	Flow m³/s	Concentration kg/m³	Particle size % > 0.062	% < 0.062
11/09/2007	Tuesday	40.5	0.197	17	83
13/09/2007	Thursday	35.1	0.171	12	88
15/09/2007	Saturday	37.0	0.195	18	82
17/09/2007	Monday	38.4	0.221	16	84
19/09/2007	Wednesday	40.5	0.243	17	83
21/09/2007	Friday	44.8	0.209	14	86
28/09/2007	Friday	26.8	0.219	11	89
01/10/2007	Monday	22.9	0.136	5	95
15/10/2007	Monday	30.9	0.102	6	94
04/11/2007	Sunday	16.7	0.093	11	89
08/12/2007	Saturday	17.3	0.098	4	96
03/03/2008	Monday	32.3	0.105	4	96
01/04/2008	Tuesday	21.1	0.094	8	92
22/04/2008	Tuesday	22.1	0.064	5	95
26/04/2008	Saturday	40.5	0.258	7	93
30/04/2008	Wednesday	40.5	0.124	9	91
07/05/2008	Wednesday	29.3	0.118	10	90
14/05/2008	Wednesday	47.2	0.116	6	94
21/05/2008	Wednesday	47.2	0.337	11	89
28/05/2008	Wednesday	47.2	0.122	4	96
04/06/2008	Wednesday	47.8	0.122	8	92
11/06/2008	Wednesday	47.8	0.165	9	91
18/06/2008	Wednesday	25.3	0.387	17	83
25/06/2008	Wednesday	30.1	0.825	19	81
02/07/2008	Wednesday	40.5	0.310	16	84
09/07/2008	Wednesday	22.1	0.570	15	85
16/07/2008	Wednesday	31.7	0.947	11	89
17/07/2008	Thursday	30.1	0.259	14	86
24/07/2008	Thursday	31.7	0.301	16	84
31/07/2008	Thursday	40.5	0.242	17	83
07/08/2008	Thursday	43.5	0.231	15	85
14/08/2008	Thursday	44.8	0.829	18	82
21/08/2008	Thursday	30.9	0.249	23	77
28/08/2008	Thursday	38.4	0.134	26	74
04/09/2008	Thursday	47.2	0.155	16	84
11/09/2008	Thursday	47.2	0.257	13	87
18/09/2008	Thursday	47.7	0.473	12	88
26/09/2008	Friday	44.8	0.118	14	86

G. Sediment inflow at RD 242

Date	Day	Location RD 242	Concentration kg/m³	Particle size % > 0.062	Particle size % < 0.062
01/03/2007	Thursday				
03/04/2007	Tuesday	6.3	0.098	10	90
21/05/2007	Monday	7.1	0.134	6	94
23/05/2007	Wednesday	5.5	0.118	12	88
25/05/2007	Friday	5.7	0.106	17	83
27/05/2007	Sunday	7.9	0.319	1	99
29/05/2007	Tuesday	4.8	0.115	0	100
31/05/2007	Thursday	7.9	0.108	3	150
02/06/2007	Saturday	7.7	0.052	0	100
04/06/2007	Monday	7.6	0.076	10	90
06/06/2007	Wednesday	7.7	0.098	4	96
08/06/2007	Friday	7.5	0.038	6	94
10/06/2007	Sunday	7.7	0.115	17	83
12/06/2007	Tuesday	7.7	0.105	5	95
14/06/2007	Thursday	7.9	0.024	5	95
16/06/2007	Saturday	7.7	0.132	8	92
20/06/2007	Wednesday	7.8	0.119	15	85
22/06/2007	Friday	7.7	0.024	14	86
24/06/2007	Sunday	7.3	0.161	8	92
26/06/2007	Tuesday	7.7	0.608	21	79
28/06/2007	Thursday	8	0.288	9	91
28/06/2007	Thursday	7.7	0.188	10	90
30/06/2007	Saturday	7.6	0.559	5	95
02/07/2007	Monday	8.1	0.240	18	82
04/07/2007	Wednesday	8.2	0.285	16	84
06/07/2007	Friday	7.9	0.219	3	97
08/07/2007	Sunday	8.0	0.130	8	92
11/07/2007	Wednesday	8.1	0.241	9	91
12/07/2007	Thursday	7.8	0.239	11	89
13/07/2007	Friday	8.1	0.481	14	86
14/07/2007	Saturday	8.0	0.343	8	92
15/07/2007	Sunday	7.3	0.272	11	89
16/07/2007	Monday	8.2	0.325	16	84
17/07/2007	Tuesday	8.3	0.337	14	86
18/07/2007	Wednesday	7.5	0.247	9	91
19/07/2007	Thursday	8.0	0.288	11	89
20/07/2007	Friday	8.3	0.474	12	88
21/07/2007	Saturday	7.2	0.293	18	82

Date	Day	Location	Concentration	Particle size	
		RD 242	kg/m^3	% > 0.062	% < 0.062
22/07/2007	Sunday	8.1	0.365	13	87
23/07/2007	Monday	8.3	0.261	13	87
24/07/2007	Tuesday	7.9	0.486	7	93
25/07/2007	Wednesday	8.2	0.274	11	89
26/07/2007	Thursday	7.7	0.330	14	86
27/07/2007	Friday	8.3	0.314	8	92
28/07/2007	Saturday	7.9	0.163	9	91
29/07/2007	Sunday	8.1	0.170	7	93
30/07/2007	Monday	6.9	0.402	11	89
31/07/2007	Tuesday	8.1	0.208	10	90
01/08/2007	Wednesday	8	0.827	13	87
02/08/2007	Thursday	8.3	0.749	13	87
03/08/2007	Friday	8.1	0.254	11	89
04/08/2007	Saturday	8.3	0.247	15	85
05/08/2007	Sunday	7.8	0.102	14	86
06/08/2007	Monday	7.1	0.193	15	85
07/08/2007	Tuesday	8.3	0.331	16	84
08/08/2007	Wednesday	7.5	0.195	20	80
09/08/2007	Thursday	8.6	0.187	15	85
10/08/2007	Friday	8.4	0.285	14	86
11/08/2007	Saturday	8.3	0.358	21	79
12/08/2007	Sunday	8.1	0.534	17	83
20/08/2007	Monday	8	0.209	23	77
22/08/2007	Wednesday	7.6	0.186	14	86
24/08/2007	Friday	7.9	0.206	15	85
26/08/2007	Sunday	7.9	0.339	12	88
28/08/2007	Tuesday	8	0.166	11	89
30/08/2007	Thursday	8	0.221	12	88
01/09/2007	Saturday	8.3	0.151	14	86
03/09/2007	Monday	8.1	0.236	12	88
05/09/2007	Wednesday	8	0.281	13	87
07/09/2007	Friday	8.2	0.220	11	89
09/09/2007	Sunday	8.3	0.195	7	93
11/09/2007	Tuesday	8.2	0.181	12	88
13/09/2007	Thursday	8.3	0.099	7	93
15/09/2007	Saturday	8.1	0.101	13	87
17/09/2007	Monday	8.3	0.080	11	89
19/09/2007	Wednesday	8.2	0.090	12	88
21/09/2007	Friday	8.1	0.097	9	91

Date	Day	Location RD 242	Concentration kg/m^3	Particle size	
				% > 0.062	% < 0.062
28/09/2007	Friday	8.3	0.115	6	94
01/10/2007	Monday	6.4	0.121	6	94
15/10/2007	Monday	6.7	0.057	7	93
04/11/2007	Sunday	6.6	0.072	9	91
08/12/2007	Saturday	6.1	0.049	5	95
03/03/2008	Monday	0	0.000	0	0
01/04/2008	Tuesday	6.5	0.108	7	93
22/04/2008	Tuesday	6.4	0.137	4	96
26/04/2008	Saturday	6	0.083	6	94
30/04/2008	Wednesday	5.7	0.109	8	92
07/05/2008	Wednesday	8.0	0.126	9	91
14/05/2008	Wednesday	7.5	0.052	5	95
21/05/2008	Wednesday	8.2	0.103	10	90
28/05/2008	Wednesday	7.8	0.274	3	97
04/06/2008	Wednesday	8.1	0.186	1	99
11/06/2008	Wednesday	8.1	0.067	2	98
18/06/2008	Wednesday	7.9	0.358	10	90
25/06/2008	Wednesday	8.3	0.147	12	88
02/07/2008	Wednesday	8.2	0.273	18	82
09/07/2008	Wednesday	8.0	0.252	17	83
16/07/2008	Wednesday	8.3	0.431	13	87
17/07/2008	Thursday	7.8	0.223	16	84
24/07/2008	Thursday	8.1	0.262	18	82
31/07/2008	Thursday	8.3	0.202	19	81
07/08/2008	Thursday	8.0	0.346	17	83
14/08/2008	Thursday	7.6	0.413	15	85
21/08/2008	Thursday	8.1	0.319	20	80
28/08/2008	Thursday	7.8	0.225	23	77
04/09/2008	Thursday	7.5	0.130	13	87
11/09/2008	Thursday	8.3	0.296	10	90
18/09/2008	Thursday	7.4	0.122	9	91
26/09/2008	Friday	8.2	0.161	11	89

H. Results of equilibrium measurements

H.1. Results of equilibrium measurements in July 2007

Canal Name	Date	Study reach	Q	TA	TW	NU*1.E-5	Area	W	DH	P
MH BRANCH	12/07/2007	1524	1646	44.8	29	22	0.096	44	23.8	1.8	25.6
MHBRANCH	16/07/2007	28680	28769	23.8	29	25	0.090	40	25.6	1.6	26.5
MH BRANCH	17/07/2007	50639	50823	15.0	29	22	0.096	34	24.7	1.4	25.3
MH BRANCH	17/07/2007	73392	73514	6.6	34	29.5	0.081	25	19.2	1.3	19.8
MA BRANCH	18/07/2007	239	361	25.0	32	23	0.094	44	22.9	1.9	24.4
MA BRANCH	18/07/2007	22652	22774	11.9	36	24	0.092	26	18.9	1.4	19.8
MA BRANCH	19/07/2007	45817	45939	5.2	33.5	24	0.092	13	11.6	1.1	12.5

	R	V	A1	A2	A3	A4	I/SI	FR	N	C*	TAU
MH BRANCH	1.7	1.0	5.24	50.35	-0.16	-1.2	533	0.242	0.03	5.79	7.051
MH BRANCH	1.5	0.6	3.08	55.99	-0.1	-1.34	876	0.152	0.034	4.6	3.779
MH BRANCH	1.3	0.4	6.43	48.71	-0.26	-1.09	5310	0.12	0.038	8.91	0.549
MH BRANCH	1.3	0.3	0.21	49.89	0	-1.07	1081	0.073	0.036	2.44	2.584
MA BRANCH	1.8	0.6	1.47	52.99	0	-1.28	3637	0.13	0.023	8.13	1.087
MA BRANCH	1.3	0.5	2	48.4	-0.04	-1.04	3077	0.122	0.028	6.96	0.947
MA BRANCH	1.0	0.4	-1.62	53.31	0.06	-1.15	2857	0.122	0.027	6.8	0.786

	U*G	TAUV	LACEYS		Bed Material Size		Fine Material Load			Bed Material Load	
			FRS	FVR	D50	Sigma	< 0.002	0.002-0.062	>0.062	D50	Sigma
MH BRANCH	0.035	7.22	5.18	1.52	0.296	1.416	6	83	87	0.26	1.82
MH BRANCH	0.020	2.25	3.57	0.58	0.131	1.364	13	97	58	0.112	1.52
MH BRANCH	0.016	0.24	1.03	0.36	0.102	1.24	33	416	22	0.102	1.20
MH BRANCH	0.010	0.68	2.93	0.13	0.198	1.375	14	188	--	0.093	1.70
MA BRANCH	0.020	0.62	1.47	0.44	0.212	1.438	13	47	--	0.084	1.30
MA BRANCH	0.016	0.43	1.48	0.38	0.130	1.864	10	80	9	0.137	1.80
MA BRANCH	0.015	0.32	1.42	0.39	0.134	1.458	20	82	17	0.171	2.13

H.2. Results of equilibrium measurements in December 2007

Canal Name	Date	Study reach	Q	TA	TW	NU*1.E-5	Area	W	DH	P
MH BRANCH	12/12/2007	1524	1646	25.5	29	22	0.140	33	21.9	1.5	23.2
MH BRANCH	13/12/2007	28680	28769	13.2	29	25	0.140	26	24.4	1.1	24.7
MH BRANCH	14/12/2007	50639	50823	7.9	29	22	0.135	34	24.4	1.4	25.0
MH BRANCH	15/12/2007	73392	73514	5.5	34	29.5	0.135	27	21.6	1.2	21.9
MA BRANCH	16/12/2007	239	361	15.2	32	23	0.142	44	23.2	1.9	24.7
MA BRANCH	17/12/2007	22652	22774	8.5	36	24	0.140	27	18.6	1.5	19.5
MA BRANCH	18/12/2007	45817	45939	4.2	33.5	24	0.137	12	11.6	1.0	12.2

Canal Name	R	V	A1	A2	A3	A4	1/S1	FR	N	C*	TAU
MH BRANCH	1.4	0.8	5.24	50.35	-0.16	-1.2	533	0.242	0.030	5.79	5.189
MH BRANCH	1.1	0.5	3.08	55.99	-0.1	-1.34	876	0.152	0.034	4.6	2.551
MH BRANCH	1.3	0.2	6.43	48.71	-0.26	-1.09	5310	0.12	0.038	8.91	0.732
MH BRANCH	1.2	0.2	0.21	49.89	0	-1.07	1081	0.073	0.036	2.44	2.110
MA BRANCH	1.8	0.3	1.47	52.99	0	-1.28	3637	0.13	0.023	8.13	0.883
MA BRANCH	1.4	0.3	2	48.4	-0.04	-1.04	3077	0.122	0.028	6.96	1.367
MA BRANCH	1.0	0.3	-1.62	53.31	0.06	-1.15	2857	0.122	0.027	6.8	0.861

Canal Name	U*G	TAUV	LACEYS	FVR	Bed Material Size		Fine Material Load		Bed Material Load		
	m/s	N/m.s	FRS		D50	Sigma	<0.002	0.002-0.062	>0.062	D50	Sigma
MH BRANCH	0.027	4.05	5.18	1.52	0.192	1.416	6	83	87	0.26	1.822
MHBRANCH	0.018	1.28	3.57	0.58	0.165	1.364	13	97	58	0.112	1.524
MH BRANCH	0.009	0.17	1.03	0.36	0.161	1.24	33	416	22	--	--
MH BRANCH	0.009	0.42	2.93	0.13	0.131	1.375	14	188	--	--	--
MA BRANCH	0.013	0.31	1.47	0.44	0.136	1.438	13	47	--	--	--
MA BRANCH	0.012	0.43	1.48	0.38	0.112	1.864	10	80	9	0.137	--
MA BRANCH	0.013	0.30	1.42	0.39	0.158	1.458	20	82	17	0.171	2.129

H.3. Results of equilibrium measurements in July 2008

Canal Name	Date	Study reach		Q	TA	TW	NU*1.E-5	Area	W	DH	P
MH BRANCH	22/07/2008	2357	2662	28.7	29	22	0.094	31	23.2	1.3	23.8
MH BRANCH	22/07/2008	27278	27583	15.7	34	29.5	0.096	26	22.9	1.2	23.5
MH BRANCH	23/07/2008	50639	50823	12.2	32	23	0.094	29	23.8	1.2	24.1
MH BRANCH	23/07/2008	73392	73514	8.1	36	24	0.096	16	17.7	0.9	18.0
MA BRANCH	27/07/2008	239	361	21.8	33.5	24	0.101	45	22.9	2.0	24.4
MA BRANCH	26/07/2008	22652	22774	11.1	33.5	24	0.101	26	19.2	1.4	20.1
MA BRANCH	26/07/2008	45817	45939	6.0	33.5	24	0.101	13	11.6	1.1	12.8

Canal Name	R	V	A1	A2	A3	A4	1/SI	FR	n	C*	TAU
MH BRANCH	1.3	0.9	6.43	48.71	-0.26	-1.09	5310	0.12	0.038	8.91	1.970
MH BRANCH	1.1	0.6	0.21	49.89	0	-1.07	1081	0.073	0.036	2.44	1.927
MH BRANCH	1.2	0.4	1.47	52.99	0	-1.28	3637	0.13	0.043	8.13	0.646
MH BRANCH	0.9	0.5	2	48.4	-0.04	-1.04	3077	0.122	0.038	6.96	1.948
MA BRANCH	1.8	0.5	-1.62	53.31	0.06	-1.15	2857	0.122	0.024	6.8	1.012
MA BRANCH	1.3	0.4	-1.62	53.31	0.06	-1.15	2857	0.122	0.028	6.8	1.733
MA BRANCH	1.0	0.5	-1.62	53.31	0.06	-1.15	2857	0.122	0.027	6.8	0.904

Canal Name	U*G	TAUV	LACEYS		Bed Material Size		Fine Material Load		Bed Material Load		
	m/s	N/m.s	FRS	FVR	D50	Sigma	<0.002	0.002-0.062	>0.062	D50	Sigma
MH BRANCH	0.033	1.83	1.03	0.36	0.081	1.24	33	416	22	--	--
MH BRANCH	0.022	1.16	2.93	0.13	0.297	1.375	14	188	--	--	--
MH BRANCH	0.015	0.27	1.47	0.44	0.248	1.438	13	47	--	--	--
MH BRANCH	0.019	0.99	1.48	0.38	0.276	1.864	10	80	9	0.137	--
MA BRANCH	0.017	0.49	1.42	0.39	0.272	1.458	20	82	17	0.171	2.129
MA BRANCH	0.015	0.72	1.42	0.39	0.272	1.458	20	82	17	0.171	2.129
MA BRANCH	0.017	0.41	1.42	0.39	0.272	1.458	20	82	17	0.171	2.129

Description of abbreviations in Appendix H.1, H.2 and H.3

Abbreviation	Description	Units	Abbreviation	Description	Units
MH Branch	Machai Branch Canal		1/S1	Hydraulic grade of selected reach	
MA Branch	Maira Branch Canal		FR	Froude No.	
Q	Water discharge	m³/s	N	Manning's n	
TA	Air temperature	°C	C*	Dimensionless Chezy coefficient	
TW	Water Temp	°C	TAU	Boundary shear stress	N/m²
NU*1.E-5	Kinematic viscosity	m²/s	U*G	MEP shear Velocity	m/s
Area	Cross-sectional area	m²	TAUV	Steam power	N/m.s
W	Top width	m	FRS	Lacey's silt factors	
DH	Hydraulic Depth	m	FVR	Flow velocity	
P	Wetted perimeter	m	D50	Median diameter	mm
R	Hydraulic radius	m	Sigma	Gradation coefficient	
V	Flow velocity	m/s	ppm	Parts per million (by weight)	
A1, A2, A3, A4	Moment coefficients of x-section	%			

Samenvatting

Net als in veel opkomende economieën en minst ontwikkelde landen is de landbouw van vitaal belang is voor de nationale economie van Pakistan. Het zorgt voor 21% van het jaarlijkse Bruto Nationaal Product (BNP), biedt werk voor 44% van de totale beroepsbevolking en draagt voor 60% bij aan de Nationale export. Pakistan heeft een totale oppervlakte van 80 Mha (miljoen hectare) met 22 Mha bouwland, waarvan 17 Mha onder irrigatie, meestal irrigatie met oppervlaktewater. Door het aride tot semi-aride klimaat, is de irrigatie in de regel nodig voor succesvolle landbouw in Pakistan.

De ontwikkeling van de moderne irrigatie in het voormalige India (Indo-Pakistan) begon in 1859 met de bouw van het Boven Bari Doab kanaal vanuit de Ravi rivier. In de loop van de tijd groeide het irrigatiesysteem van Pakistan uit tot 's werelds grootste aaneengesloten irrigatie systeem onder natuurlijk verval, bekend als het Indus Basin Irrigatie Systeem (IBIS). In het IBIS worden bijna alle irrigatie kanalen rechtstreeks gevoed vanuit rivieren, terwijl de rivier afvoeren grote ladingen sediment bevatten. Irrigatie kanalen die hiermee worden gevoed ontvangen enorme hoeveelheden sedimenten, die vervolgens, afhankelijk van de hydrodynamische condities van de kanalen, in de irrigatie kanalen worden afgezet. Sediment afzetting in irrigatie kanalen veroorzaakt ernstige problemen bij beheer en onderhoud. Studies tonen aan dat het afgezette sediment kan leiden tot een afname met 40% van de beschikbare afvoercapaciteit van de irrigatie kanalen.

Onderzoekers streven er sinds vele jaren naar om dit probleem op een duurzame manier te beheersen en een aantal benaderingen zijn in dit verband ingevoerd. Als een eerste stap worden de sedimenten bij de inname punten vanuit de rivieren gecontroleerd met kunstwerken die zoveel mogelijk voorkomen dat sediment in de kanalen komt en kunstwerken die het binnengekomen sediment uit de kanalen afleiden. Dan wordt een zodanige ontwerp aanpak voor de kanalen gevolgd dat de sedimenten zoveel mogelijk in suspensie blijven, teneinde ze te verdelen over de geïrrigeerde velden. Zelfs dan hebben de sedimenten de neiging om zich in de irrigatie kanalen af te zetten wat tot ernstige problemen bij beheer en onderhoud kan leiden, wat vervolgens leidt tot regelmatige opschoning campagnes om ervoor te zorgen dat het water in de kanalen blijft stromen. Dit veroorzaakt een continue belasting voor de nationale economie. In opkomende economieën en minst ontwikkelde landen is adequate en tijdige beschikbaarheid van middelen voor beheer en onderhoud over het algemeen een probleem. Dit veroorzaakt vertraging in het onderhoud van de kanalen, waardoor hun hydraulische capaciteit wordt beïnvloed. Water wordt dan onvoldoende en ongelijkmatig aan de water gebruikers geleverd.

Het verhaal wordt nog ingewikkelder als het gaat om op benedenstroomse vraag gestuurde irrigatie kanalen onder flexibele bediening. Bij beheer op basis van vaste aanvoer is in de kanalen altijd dezelfde stroomsnelheid en een dergelijk beheer leidt door de voldoende snelheden in de regel niet tot sediment afzetting in het kanaal profiel. Echter, bij de vraag op basis van flexibele bediening lopen de kanalen niet altijd op volle capaciteit maar zijn ze onderhevig aan wisselende stroomsnelheden, die worden bepaald door het groeistadium van de gewassen en de daarbij benodigde hoeveelheid water in het door het kanaal te bedienen gebied. Een dergelijke wijze van kanaal beheer is niet altijd gunstig is voor sediment transport bij geringe water aanvoer, de stroomsnelheden worden gering en er kan depositie van sediment in het kanaal profiel optreden. Hier rijzen de vragen wat voor soort hydrodynamische relaties sediment afzetting in de op benedenstroomse vraag gestuurde irrigatie kanalen kunnen voorkomen en hoe deze relaties kunnen worden toegepast, terwijl wordt voorzien in de voor de gewassen benodigde hoeveelheden water in het te bedienen gebied? Hoe kan de behoefte aan onderhoud worden geminimaliseerd door beter kanaal beheer met betrekking tot sediment transport?

De studie is opgezet om dergelijke typen van relaties en praktijken te onderzoeken om sediment transport in op benedenstroomse vraag gestuurde irrigatie kanalen zodanig te beheren teneinde een maximale hydraulische efficiëntie bij een minimum aan onderhoud te bereiken. De hypothese van de studie stelt dat in op basis van vraag gestuurde irrigatie kanalen de omvang van de depositie van sediment kan worden geminimaliseerd en zelfs dat de sedimenten die tijdens de lage vraag naar water door de gewassen zijn gedeponeerd, tijdens piek perioden van water aanvoer weer in suspensie kunnen komen. Op deze manier kan een evenwicht worden gehandhaafd in sediment depositie en resuspensie door adequaat kanaal beheer.

In deze studie zijn twee computer modellen gebruikt, namelijk Simulatie van Irrigatie Kanalen (SIC) en Sediment Transport in Irrigatie Kanalen (SETRIC). Beide modellen zijn eendimensionaal en kunnen stationaire en niet stationaire afvoeren simuleren (SETRIC alleen stationaire afvoeren), alsmede niet stationair sediment transport in irrigatie kanalen. Het SIC model heeft de mogelijkheid om sediment transport te simuleren onder niet-stationaire stroming situaties en kan toegepast worden om het effect van sediment depositie op de hydraulische prestaties van irrigatie kanalen te simuleren. Daarentegen heeft het SETRIC model het voordeel dat, rekening wordt gehouden met de ontwikkeling van ribbels in de bodem en hun effect op de weerstand tegen stroming, wat een kritische factor is in het ontwerp en beheer van irrigatie kanalen. Voor het SETRIC model is een nieuwe module met betrekking tot sediment transport simulaties in op benedenstroomse vraag gestuurde irrigatie kanalen opgenomen.

De studie is uitgevoerd voor het Boven Swat kanaal - Pehure Hoog Niveau kanaal (USC-PHLC) irrigatie systeem, dat bestaat uit drie kanalen, Machai Branch kanaal, PHLC en Maira Branch kanaal. Het Machai Branch kanaal kent bovenstrooms gestuurde aanvoer en de twee andere kanalen zijn op benedenstroomse vraag gestuurd. Deze kanalen zijn onderling verbonden. De PHLC en Machai Branch kanalen voeden het Maira Branch kanaal, alsmede hun eigen irrigatie systemen. PHLC ontvangt water uit het Tarbela reservoir en het Machai Branch kanaal van de Swat rivier via het USC. Water vanuit het Tarbela reservoir is op dit moment sediment vrij, terwijl het water uit de Swat rivier sediment bevat. Verschillende studies hebben echter aangegeven dat het Tarbela reservoir binnenkort zal zijn gevuld met sediment en zich zal gedragen als een door een rivier gevoed systeem. Dan ontvangt ook PHLC water dat sedimenten bevat. De ontwerp afvoeren van de Machai, PHLC en Maira Branch kanalen zijn respectievelijk 65, 28 en 27 m^3/s. Het te bedienen gebied van het USC-PHLC irrigatie systeem is 115.800 ha.

Het USC-PHLC irrigatie systeem is onlangs gerenoveerd waarbij de capaciteit is verhoogd van 0,34 l/s/h tot 0,67 l/s/h. Het bovenstroomse USC systeem, vanaf de inlaat tot de Machai Branch RD 242 (het controle punt waar het op benedenstroomse vraag gestuurde systeem start) werd in 1995 gerenoveerd, terwijl het systeem benedenstrooms van RD 242 in 2003 werd gerenoveerd. Het bovenstroomse gedeelte van het Machai Branch kanaal tot een punt op ongeveer 74.000 m wordt beheerd op basis van een vaste aanvoer, terwijl het benedenstroomse deel van het Machai Branch kanaal, Maira Branch kanaal en het PHLC wordt beheerd op basis van gedeeltelijk vraag gestuurde flexibele aanvoer. Het gedeeltelijk vraag gestuurde systeem wordt beheerd op basis van de waterbehoefte van de gewassen en volgt een op gewas gebaseerd irrigatie beheer (CBIO) schema. Wanneer de vraag naar water door de gewassen tot minder dan 80% van het volledige aanbod daalt, wordt onder de secundaire verdeelwerken een roulatiesysteem ingevoerd. Tijdens perioden met een zeer geringe behoefte aan water voor de gewassen wordt, op basis van de uitgangspunten voor het beheer van het kanaal de wateraanvoer niet verder verminderd dan tot een minimum van 50% van de volledige aanvoer.

Het onderzoek bestond uit twee jaar veldwerk, waarbij dagelijkse gegevens zijn verzameld betreffende het beheer van de kanalen, maandelijkse aanvoer van sediment gedurende perioden met een lage aanvoer van sediment en wekelijkse gegevens in perioden met piekconcentraties. Er zijn drie massabalans studies uitgevoerd waarbij de water en sediment aanvoer en afvoer zijn bemeten, tevens is op geselecteerde locaties langs het kanaal sediment in suspensie bemonsterd en ook zijn in de secondaire kanalen, onmiddellijk benedenstrooms van de verdeelwerken mengmonsters genomen. Voorts zijn in de vier maanden tijdens de piek sediment seizoen juni, juli, augustus en september, massabalans studies uitgevoerd op basis van mengmonsters van sediment teneinde de water en sediment aanvoer naar en afvoer vanuit het systeem te bepalen. Teneinde het effect van sediment transport op de morfologie van de kanalen te bepalen zijn in vijf dwarsdoorsneden peilingen uitgevoerd op basis waarvan de veranderingen in het bodemprofiel zijn gemeten. Op basis van deze gegevens zijn de twee computer modellen, die in deze studie zijn gebruikt, gekalibreerd en gevalideerd voor stroming en sediment transport simulaties.

De op benedenstroomse vraag gestuurde component van het systeem wordt automatisch gecontroleerd en de PHLC is uitgerust met het Supervisory Control and Data Acquisition (SCADA) systeem op de inlaatwerken. Elke waterinlaat of blokkade door de Vereniging van Watergebruikers via de secundaire kanalen, of elke variatie in de aanvoer vanuit het Machai Branch kanaal wordt automatisch aangepast door het SCADA systeem op het Gandaf afvoerpunt, het PHLC verdeelwerk. Het SCADA systeem heeft Proportionele Integrale (PI) afvoer controle apparatuur. De studie wees uit dat de bestaande PI coëfficiënten leidden tot vertraging in de afvoeren en resulteerden in een lange tijd voordat de afvoeren stabiel werden. De afvoeren toonden een oscillerend gedrag dat de werking van de hydro-mechanisch bediende op benedenstroomse vraag gestuurde doorstroombare radiale schuiven (AVIO) en overstroombare radiale schuiven (AVIS) schuiven beïnvloedde. Na de kalibratie en validatie van het model is de PI controle apparatuur afgestemd en is een verbeterd beheer van het kanaal voorgesteld, wat zou helpen om de duurzaamheid van het systeem te vergroten en de operationele efficiëntie van de kanalen te verbeteren.

Uit de veldgegevens bleek dat gedurende de studieperiode de sedimentatie in de bestudeerde irrigatie kanalen binnen de controle grenzen bleef. De inkomende hoeveelheden sediment waren over het algemeen lager dan de capaciteit met betrekking tot sediment transport van de bestudeerde irrigatie kanalen. Dientengevolge werden de aangevoerde hoeveelheden sediment door de hoofd kanalen getransporteerd en doorgevoerd naar de secondaire kanalen. De sediment transport capaciteiten van de bestudeerde irrigatie kanalen werden berekend voor stationaire en niet-stationaire situaties. Uit gegevens betreffende het beheer van de kanalen bleek dat het systeem meer werd beheerd op basis van een op aanvoer gebaseerd beheer (SBO) dan op CBIO. Uit de morfologische gegevens bleek dat er geen significante depositie in de bestudeerde kanalen optrad. Daarom was er geen bijzondere invloed op de werking van de kanalen en op de hydraulische efficiëntie die werd toegeschreven aan het sediment transport.

Zoals eerder is aangegeven zal het Tarbela reservoir binnenkort met sedimenten gevuld zijn en zal PHLC sediment rijk water uit het reservoir ontvangen. Er zijn verschillende studies in beschouwing genomen betreffende het tijdstip waarop het sediment rijke water in het PHLC zal stromen en wat de kenmerken en concentraties van het naar PHLC aangevoerde water vanuit het reservoir zullen zijn. De studies wijzen uit dat het aangevoerde sediment vanuit het Tarbela reservoir veel meer zal zijn dan de sediment transport capaciteit van het PHLC en het Maira Branch kanaal onder maximale wateraanvoer aan zullen kunnen. Dit scenario zal leiden tot sediment transport problemen in de op benedenstroomse vraag gestuurde kanalen, met name wanneer zij zullen worden beheerd op basis van CBIO.

Diverse opties voor het beheer zijn gesimuleerd en worden gepresenteerd om een beter beheer van de sedimenten in de bestudeerde kanalen bij toevoer van sediment vanuit het Tarbela reservoir. De hydraulische prestaties van de op benedenstroomse vraag gestuurde kanalen zullen bij dit scenario afnemen en frequent onderhoud en reparatie zullen nodig zijn om de kanalen te onderhouden. Verschillende opties zijn onderzocht hoe om te gaan met het probleem. De studie presenteert een sediment beheersplan voor op benedenstroomse vraag gestuurde irrigatie kanalen met verbeteringen in kanaal ontwerp en beheer in combinatie met de aanleg van sedimentatie bekkens bij de verdeelwerken.

Momenteel is sedimentatie in de irrigatie kanalen van dit onderzoek geen groot probleem voor beheer en onderhoud. Het zal zich echter tot een groot probleem ontwikkelen als de aanvoer van sediment vanuit het Tarbela reservoir begint. De onderhoud kosten voor de kanalen zullen de pan uit rijzen en de hydrodynamische eigenschappen van deze kanalen zullen ook worden beïnvloed. In dit onderzoek zijn een aantal mogelijkheden geëvalueerd en voorgesteld teneinde het te verwachten probleem van het sediment transport in deze irrigatie kanalen te kunnen opvangen, om hun hydraulische eigenschappen op het gewenste niveau te houden en om de onderhoud kosten te minimaliseren. De eerste en de belangrijkste gevolgen van sediment depositie zal vermindering van de afvoercapaciteiten van de kanalen zijn wat zal resulteren in verhoging van de waterpeilen. De verhoging van waterpeilen zal leiden tot een vermindering in de wateraanvoer naar de kanalen wat te wijten zal zijn aan de automatische regeling van de afvoeren. Dit kan worden gerealiseerd door een tijdelijke en beperkte verhoging in de na te streven waterpeilen in relatie tot het maximale energie verlies bij de op benedenstroomse vraag gestuurde AVIS/AVIO stuw. Voorts, om het effect van het verhoogde waterpeil op de afvoer te minimaliseren door de AVIS/AVIO radiale schuiven, kan het peilverschil in deze kanalen relatief klein worden gehouden, om de radiale schuiven minder gevoelig te maken voor veranderingen in het niveau van het water. Tevens moeten voor een efficiënte toevoer van sediment naar de secundaire kanalen, moeten de secundaire verdeelwerken aan de benedenstroomse kant in de buurt van AVIS/AVIO stuwen worden gelokaliseerd. Meer sediment zal worden geloosd, omdat de turbulente menging van sediment aan de benedenstroomse kant van de verdeelwerken meer sediment in suspensie houdt. Bovendien moeten, teneinde depositie te minimaliseren, tijdens perioden met een piek concentratie aan sediment, de kanalen worden beheerd op basis van aanvoer.

Sediment transport in het algemeen en in irrigatie kanalen, in het bijzonder, is een van de meest bestudeerde en besproken onderwerpen op het gebied van de stromingsleer over de hele wereld. In verband met het ontwerpen en beheren van irrigatie kanalen die worden gevoed met sediment rijk water is het ook uitgebreid bestudeerd in het stroomgebied van de Indus rivier. De uitkomsten van de regime theorie van Lacey en de daarop volgende werkzaamheden zijn het resultaat van deze studies. Naast de regime methode zijn diverse andere methoden voor een stabiel ontwerp van kanalen ontwikkeld, zoals de methode van de toegestane snelheid, de schuifspanning methode, de rationele methode, enz. Hoe dan ook, het is een feit dat het beheer van sediment transport in irrigatie kanalen nog steeds een uitdagende taak is, zelfs na al deze onderzoeken en studies. Omdat de meeste kennis over sediment transport empirisch van aard is, hebben de meeste sediment transport formules een ingebouwde willekeur, waardoor voorspellingen moeilijk zijn, als de omstandigheden veranderen. Het vereist veel zorg om een sediment transport formule, die is ontwikkeld onder een reeks voorwaarden, in andere situaties toe te passen. Daarom is het uiterst belangrijk om de oorsprong van de ontwikkeling van de formules en de beperkingen in verband met de toepassing te begrijpen voordat de formules onder andere omstandigheden worden toegepast. De invoering van numerieke modellering maakt het betrekkelijk eenvoudig om sediment transport relaties te testen en onder lokale omstandigheden vorm te geven door het uitvoeren van een aantal simulaties en het kalibreren van de formule in relatie tot veld metingen. De voorspellingen met betrekking tot sediment transport kunnen op deze wijze betrouwbaar worden gemaakt en gebruikt voor verdere analyse.

About the author

Sarfraz Munir was born in Narowal, Pakistan in 1974. He received his Bachelor and Master degrees in Agricultural Water Management from the North Western Frontier Province (now Khyber Pakhtunkhwa) Agricultural University Peshawar, Pakistan in 1997 and 2000 respectively. He served for about 10 years from 1999 to 2009 in the International Water Management Institute (IWMI) in Pakistan. He has been engaged in a number of projects on performance assessment of irrigation systems in Pakistan, agricultural water management, environmental impact assessment of wastewater irrigation and the evaluation of modernized irrigation systems. His key assignments included field measurements for agricultural and environmental water management like flow measurements in irrigation canals, hydraulic structures' and downstream gauges' calibration, longitudinal and cross-sectional surveys in irrigation canals, groundwater observations, installation of necessary equipment for surface water and groundwater observations, *in situ* water quality measurements, laboratory analysis of plant, soil and water quality, socio-economic surveys, hydraulic modelling, synthesis of field data and report writing. From 2006 to 2011 he conducted his PhD study in Hydraulic Engineering - Land and Water Development Core, UNESCO-IHE Institute for Water Education Delft, the Netherlands.

T - #0141 - 160425 - C286 - 244/170/15 - PB - 9780415669474 - Gloss Lamination